AF388665

Carnot Cycle and Heat Engine Fundamentals and Applications

Special Issue Editor

Michel Feidt

MDPI • Basel • Beijing • Wuhan • Barcelona • Belgrade • Manchester • Tokyo • Cluj • Tianjin

Special Issue Editor
Michel Feidt
Université de Lorraine
France

Editorial Office
MDPI
St. Alban-Anlage 66
4052 Basel, Switzerland

This is a reprint of articles from the Special Issue published online in the open access journal *Entropy* (ISSN 1099-4300) (available at: https://www.mdpi.com/journal/entropy/special_issues/ carnot_cycle).

For citation purposes, cite each article independently as indicated on the article page online and as indicated below:

LastName, A.A.; LastName, B.B.; LastName, C.C. Article Title. *Journal Name* **Year**, *Article Number*, Page Range.

ISBN 978-3-03928-845-8 (Pbk)
ISBN 978-3-03928-846-5 (PDF)

Contents

About the Special Issue Editor

Michel Feidt, emeritus professor at the University of Lorraine France, where he has spent his entire career in education and research. His main interests are thermodynamics and energy. He is a specialist of infinite physical dimensions optimal thermodynamics (FDOT) from a fundamental point of view, illustrating the necessity to consider irreversibility to optimize systems and processes and characterize upper bound efficiencies. He has published many articles in journals and books: more than 120 papers and more than 5 books. He participates actively in numerous international and national conferences on the same subject. He has developed 55 final contracts reports and was the director of 43 theses. He has been member of more than 110 doctoral committees. He is a member of the scientific committee of more than 5 scientific journals and editor-in-chief of one journal.

Editorial

Carnot Cycle and Heat Engine: Fundamentals and Applications

Michel Feidt

Laboratory of Energetics, Theoretical and Applied Mechanics (LEMTA), URA CNRS 7563, University of Lorraine, 54518 Vandoeuvre-lès-Nancy, France; michel.feidt@univ-lorraine.fr

Received: 6 March 2020; Accepted: 13 March 2020; Published: 18 March 2020

After two years of exchange, this specific issue dedicated to the Carnot cycle and thermomechanical engines has been completed with ten papers including this editorial.

Our thanks are extended to all the authors for the interesting points of view they have proposed: this issue confirms the strong interactions between fundamentals and applications in thermodynamics.

Regarding the list of published papers annexed at the end of this editorial, it appears that six papers [1–6] are basically concerned with thermodynamics concepts. The three others [7–9] employ thermodynamics for specific applications. Only the papers included in the special issue are cited and analyzed in this editorial.

In "OTEC Maximum Net Power Output Using Carnot Cycle and Application to Simplify Heat Exchanger Selection", Fontaine et al. [7] consider the maximum net power for the OTEC system using the Carnot cycle to simplify heat exchanger selection. The paper "Energy and Exergy Evaluation of a Two-Stage Axial Vapor Compressor on the LNG Carrier", by Poljak et al. [8], particularizes energy and exergy evaluation to a component of a system: a two-stage axial vapor compressor for LNG application. In "Second Law Analysis for the Experimental Performances of a Cold Heat Exchanger of a Stirling Refrigeration Machine", Djetel-Gothe et al. [9] also use second law analysis to quantify experimental performances of a cold heat exchanger of a Stirling refrigerator system.

The first conclusion is that we must enlarge the subject from engines to reverse cycle machines, and particularly those dedicated to low temperature applications. This is confirmed by the paper "Global Efficiency of Heat Engines and Heat Pumps with Non-Linear Boundary Conditions" [3] which is concerned both with efficiency of heat pumps and heat engines. This paper establishes the connection between applications and concepts, mainly global efficiency in this case. This concept is fruitful due to its non-dimensional form. Lundqvist and Öhman [3] use a black box method to compare thermal efficiencies of different scale and type of engines and heat pumps. The influence of boundary conditions is exemplified. Using FTT (Finite Time Thermodynamics) and Max power cycle approaches easily enable a black box modeling with various (linear or not) boundary conditions.

Two papers give respectively a short history regarding efficiency at maximum power and an analysis of the methodology used in modeling and optimization of Carnot engines [2,6]. It appears that in any of the cases specific physical dimensions are correlated to the choice of objective function and constraints. This is why we preconize the acronym FDOT (Finite physical Dimensions Optimal Thermodynamics) instead of FTT. In "Progress in Carnot and Chambadal Modeling of Thermomechanical Engines by Considering Entropy Production and Heat Transfer Entropy", Feidt and Costea [6] details some progress in Carnot and Chambadal modeling of thermomechanical engines by considering entropy production as well as heat transfer entropy.

The paper by Gonzalez-Ayala et al. [1] has the same philosophy as the paper by Feidt and Costea [6] but using finite time heat engine models compared to low dissipation models. The proposed models take account of heat leak and internal irreversibilities related to time; the maximum power (MP) regime is covered. Upper and lower bounds of MP efficiency are reported depending on the heat transfer law.

The paper by Xue and Guo [4] is more exotic. It re-examines the Clausius statement of the second law of thermodynamics. This paper introduces an average temperature method.

We hope to have the opportunity to continue to report on the progress of the subject in the near future, extending the subject to reverse cycle configurations.

Acknowledgments: We express our thanks to the authors of the above contributions, and to the journal Entropy and MDPI for their support during this work.

Conflicts of Interest: The author declares no conflict of interest.

References

1. Gonzalez-Ayala, J.; Roco, J.M.M.; Medina, A.; Calvo Hernández, A. Carnot-Like Heat Engines Versus Low-Dissipation Models. *Entropy* **2017**, *19*, 182. [CrossRef]
2. Feidt, M. The History and Perspectives of Efficiency at Maximum Power of the Carnot Engine. *Entropy* **2017**, *19*, 369. [CrossRef]
3. Lundqvist, P.; Öhman, H. Global Efficiency of Heat Engines and Heat Pumps with Non-Linear Boundary Conditions. *Entropy* **2017**, *19*, 394. [CrossRef]
4. Xue, T.-W.; Guo, Z.-Y. What Is the Real Clausius Statement of the Second Law of Thermodynamics? *Entropy* **2019**, *21*, 926. [CrossRef]
5. Chimal-Eguia, J.C.; Paez-Hernandez, R.; Ladino-Luna, D.; Velázquez-Arcos, J.M. Performance of a Simple Energetic-Converting Reaction Model Using Linear Irreversible Thermodynamics. *Entropy* **2019**, *21*, 1030. [CrossRef]
6. Feidt, M.; Costea, M. Progress in Carnot and Chambadal Modeling of Thermomechanical Engine by Considering Entropy Production and Heat Transfer Entropy. *Entropy* **2019**, *21*, 1232. [CrossRef]
7. Fontaine, K.; Yasunaga, T.; Ikegami, Y. OTEC Maximum Net Power Output Using Carnot Cycle and Application to Simplify Heat Exchanger Selection. *Entropy* **2019**, *21*, 1143. [CrossRef]
8. Poljak, I.; Orović, J.; Mrzljak, V.; Bernečić, D. Energy and Exergy Evaluation of a Two-Stage Axial Vapour Compressor on the LNG Carrier. *Entropy* **2020**, *22*, 115. [CrossRef]
9. Djetel-Gothe, S.; Lanzetta, F.; Bégot, S. Second Law Analysis for the Experimental Performances of a Cold Heat Exchanger of a Stirling Refrigeration Machine. *Entropy* **2020**, *22*, 215. [CrossRef]

Article

Carnot-Like Heat Engines Versus Low-Dissipation Models

Julian Gonzalez-Ayala [1,*], José Miguel M. Roco [1,2], Alejandro Medina [1] and Antonio Calvo Hernández [1,2,*]

[1] Departamento de Física Aplicada, Universidad de Salamanca, 37008 Salamanca, Spain; roco@usal.es (J.M.M.R.); amd385@usal.es (A.M.)

[2] Instituto Universitario de Física Fundamental y Matemáticas (IUFFyM), Universidad de Salamanca, 37008 Salamanca, Spain

* Correspondence: jgonzalezayala@usal.es (J.G.-A.); anca@usal.es (A.C.H.)

Academic Editor: Michel Feidt

Received: 20 March 2017; Accepted: 20 April 2017; Published: 23 April 2017

Abstract: In this paper, a comparison between two well-known finite time heat engine models is presented: the Carnot-like heat engine based on specific heat transfer laws between the cyclic system and the external heat baths and the Low-Dissipation model where irreversibilities are taken into account by explicit entropy generation laws. We analyze the mathematical relation between the natural variables of both models and from this the resulting thermodynamic implications. Among them, particular emphasis has been placed on the physical consistency between the heat leak and time evolution on the one side, and between parabolic and loop-like behaviors of the parametric power-efficiency plots. A detailed analysis for different heat transfer laws in the Carnot-like model in terms of the maximum power efficiencies given by the Low-Dissipation model is also presented.

Keywords: thermodynamics; optimization; entropy analysis

---

1. Introduction

A cornerstone in thermodynamics is the analysis of the performance of heat devices. Since the Carnot's result about the maximum possible efficiency that any heat converter operating between two heat reservoirs might reach, the work in this field is mainly focused on how to fit real-life devices as close as possible to the main requirement behind the Carnot efficiency value, i.e., the existence of infinite-time, quasi-static processes. However, real-life devices work under finite-time and finite-size constraints, thus giving finite power output. Over the last several decades, one of the most popular models in the physics literature to analyze finite-time and finite-size heat devices has been the so-called Carnot-like model. Inspired by the work reported by Curzon–Ahlborn (CA) [1], this model provides a first valuable approach to the behavior of real heat engines.

In this model, it is assumed an internally reversible Carnot cycle coupled irreversibly with two external thermal reservoirs (endoreversible hypothesis) through some heat transfer laws and some phenomenological conductances related with the nature of the heat fluxes and the properties of the materials and devices involved in the transport phenomena. Without any doubt, the main result was the so-called CA-efficiency $\eta = 1 - \sqrt{\tau}$ (where $\tau = T_c/T_h$ is the ratio of the external cold and hot heat reservoirs). It accounts for the efficiency at maximum–power (MP) conditions when the heat transfer laws are considered linear with the temperature difference between the external heat baths and the internal temperatures of the isothermal processes at which the heat absorption and rejection occurs. Later extensions of this model included the existence of a heat leak between the two external baths and the addition of irreversibilities inside of the internal cycle. With only three main ingredients (heat leak, external coupling, and internal irreversibilities) the Carnot-like model has been used as a paradigmatic model to confront many research results coming from macroscopic, mesoscopic and microscopic fields [1–28]. Particularly relevant have been those results concerned with the optimization not only of the power output but also of different thermodynamic and/or thermo-economic figures of merit and additionally the analysis on the universality of the efficiency at MP (or on other figures of merit as ecological type [29–34]).

Complementary to the CA-model, the Low-Dissipation (LD) model, proposed by Esposito et al. in 2009 [35], consists of a Carnot engine with small deviations from the reversible cycle through dissipations located at the isothermal branches which occur at finite-time. The nature of the dissipations (entropy generation) are encompassed in some generic dissipative coefficients, so that the optimization of power output (or any other figure of merit) is made easily through the contact times of the engine with the hot and cold reservoirs [36–39]. In this way, depending on the symmetry of the dissipative coefficients, it is possible to recover several results of the CA-model. In particular, the CA-efficiency is recovered in the LD-model under the assumption of symmetric dissipation. Recently, a description of the LD model in terms of characteristic dimensionless variables was proposed in [40–42]. From this treatment, it is possible to separate efficiency-power behaviors typical of CA-endoreversible engines as well as irreversible engines according to the imposed time constraints. If partial contact times are constrained, then one obtains open parabolic (endoreversible) curves; otherwise, if total time is fixed, one obtains closed loop-like curves.

The objective of this paper is to analyze in which way the Carnot-like heat engines (dependent on heat transfer laws) and the LD models (dependent on a specific entropy generation law) are related and how the variables of each one are connected. This allows for an interpretation of the heat transfer laws, including the heat leak, in terms of the bounds for the efficiency at MP provided by the LD-model, which, in turn, are dependent on the relative symmetries of the dissipations constants and the partial contact times.

The article is organized as follows: in Section 2, a correspondence among the variables of the two models for heat engines (HE) is made. In Section 3, we study the region of physically acceptable values for the Carnot-like HE variables depending on the heat leak. In Section 4, the study of the MP regime is made, showing that a variety of results between both descriptions can be recovered only in a certain range of heat transfer laws; in particular, we analyze the efficiency vs. power curves behaviors. Finally, some concluding remarks are presented in Section 5.

2. Correspondence between the HE's Variables of Both Models

A key point to establish the linkage between both models is the entropy production. By equaling the entropy change stemming from both frameworks it is possible to give the relations among the variables that describe each model (see Figure 1).

In the LD case (see Figure 1a), the base-line Carnot cycle works between the temperatures T_c and $T_h > T_c$, the entropy change along the hot (cold) isothermal path is $\Delta S = -\frac{Q_h}{T_h}$ ($\Delta S = +\frac{Q_h}{T_c}$) and the times to complete each isotherm processes are t_h and t_c, respectively. The adiabatic processes, as usual, are considered as instantaneous, though the influence of finite adiabatic times has been reported in the LD-model in [43]. The deviation from the reversible scenario in the LD approximation is considered by an additional contribution to the entropy change at the hot and cold reservoirs given by

$$\Delta S_{T_h} = -\Delta S + \frac{\Sigma_h}{t_h}, \tag{1}$$

$$\Delta S_{T_c} = \Delta S + \frac{\Sigma_c}{t_c}, \tag{2}$$

where Σ_h and Σ_c are the so-called dissipative coefficients. The signs $-(+)$ take into account the direction of the heat fluxes from (toward) the hot (cold) reservoir in such a way that Q_c and Q_h are positive quantities. Then, the total entropy generation is given by

$$\Delta S_{tot} = \frac{\Sigma_h}{t_h} + \frac{\Sigma_c}{t_c}. \tag{3}$$

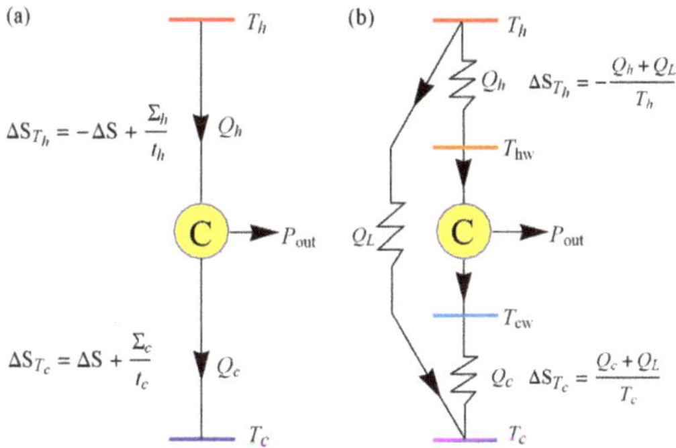

Figure 1. (a) Sketch of a low dissipation heat engine characterized by entropy generation laws ΔS_{T_h} and ΔS_{T_c}; (b) Sketch of an irreversible Carnot-like heat engine characterized by generic heat transfers Q_h, Q_c and Q_L.

At this point, it is helpful to use the dimensionless variables defined in [40]: $\alpha \equiv t_c/t$, $\tilde{\Sigma}_c \equiv \Sigma_c/\Sigma_T$ and $\tilde{t} \equiv (t\,\Delta S)/\Sigma_T$, where $t = t_h + t_c$ and $\Sigma_T \equiv \Sigma_h + \Sigma_c$. In this way, it is possible to define a characteristic total entropy production per unit time for the LD-model as

$$\widetilde{\Delta S}_{tot} \equiv \frac{\Delta S_{tot}}{\tilde{t}\,\Delta S} = \frac{\Delta S_{tot}}{t} \frac{\Sigma_T}{\Delta S^2} = \frac{1}{\tilde{t}} \left[\frac{1 - \tilde{\Sigma}_c}{(1-\alpha)\tilde{t}} + \frac{\tilde{\Sigma}_c}{\alpha \tilde{t}} \right]. \tag{4}$$

In the irreversible Carnot-like HE, the entropy generation of the internal reversible cycle is zero and the total entropy production is that generated at the external heat reservoirs (see Figure 1b). By considering the same sign convention as in the LD model $Q_h = T_{hw}\,\Delta S \geq 0$ and $Q_c = T_{cw}\,\Delta S \geq 0$, where ΔS is the entropy change in the hot isothermal branch of the reversible Carnot cycle, and a heat leak $Q_L \geq 0$ between the reservoirs T_h and T_c, then

$$\Delta S_{T_h} = -\frac{Q_h}{T_h} - \frac{Q_L}{T_h} = -\Delta S + \left(1 - a_h^{-1} - \tau \frac{Q_L}{T_c \Delta S}\right) \Delta S, \tag{5}$$

$$\Delta S_{T_c} = \frac{Q_c}{T_c} + \frac{Q_L}{T_c} = \Delta S + \left(a_c - 1 + \frac{Q_L}{T_c \Delta S}\right) \Delta S, \tag{6}$$

where $a_h = T_h/T_{hw} \geq 1$ and $a_c = T_{cw}/T_c \geq 1$. By introducing a characteristic heat leak $\tilde{Q}_L \equiv Q_L/(T_c \Delta S)$ and a comparison with Equations (1) and (2) gives the expressions associated with the dissipations

$$\frac{\Sigma_h}{t_h} = \left(1 - a_h^{-1} - \tau \tilde{Q}_L\right) \Delta S, \tag{7}$$

$$\frac{\Sigma_c}{t_c} = \left(a_c - 1 + \tilde{Q}_L\right) \Delta S. \tag{8}$$

By assuming that the ratio $t_c/(t_c + t_h)$ is the same in both descriptions, then we introduce $\alpha = 1/(1 + t_h/t_c)$ into the Carnot-like model, and $\widetilde{\Sigma}_c = \Sigma_c/\Sigma_T$ and $\widetilde{t} = t\Delta S/\Sigma_T$ are

$$\widetilde{\Sigma}_c^{-1} = 1 + \left(\frac{1-\alpha}{\alpha}\right)\left(\frac{1 - a_h^{-1} - \tau\widetilde{Q}_L}{a_c - 1 + \widetilde{Q}_L}\right),\tag{9}$$

$$\widetilde{t} = \frac{1}{\alpha\left(a_c - 1 + \widetilde{Q}_L\right)\left[1 + \left(\frac{1-\alpha}{\alpha}\right)\left(\frac{1 - a_h^{-1} - \tau\widetilde{Q}_L}{a_c - 1 + \widetilde{Q}_L}\right)\right]},\tag{10}$$

which are the relations between the characteristic variables of the LD model and the variables of the Carnot-like HE. This is summarized in the following expression:

$$\frac{\widetilde{\Sigma}_c}{\alpha\widetilde{t}} = a_c - 1 + \widetilde{Q}_L.\tag{11}$$

3. Physical Space of the HE Variables

We stress that all the above results between variables hold for arbitrary heat transfer laws in the Carnot-like model. As a consequence, above equations provide the generic linkage between both descriptions, and, from them, useful thermodynamic information can be extracted.

In Figure 2a, the internal temperatures for the irreversible Carnot-like HE, contained in a_h and a_c, are depicted with fixed values $\tau = 0.2$, $\alpha = 0.2$ and $\widetilde{\Sigma}_c = 0.5$. Notice that, in order to obtain thermal equilibrium between the auxiliary reservoirs and the external baths (i.e., to achieve the reversible limit), it is necessary that $Q_L = 0$. As soon as a heat leak appears, $T_{hw} < T_h$, meanwhile $T_{cw} = T_c$ is always a possible configuration. As the heat leak increases in the HE, the internal temperatures get closer to each other until the limiting situation where $T_{hw} = T_{cw}$ (see contact edge in Figure 2a).

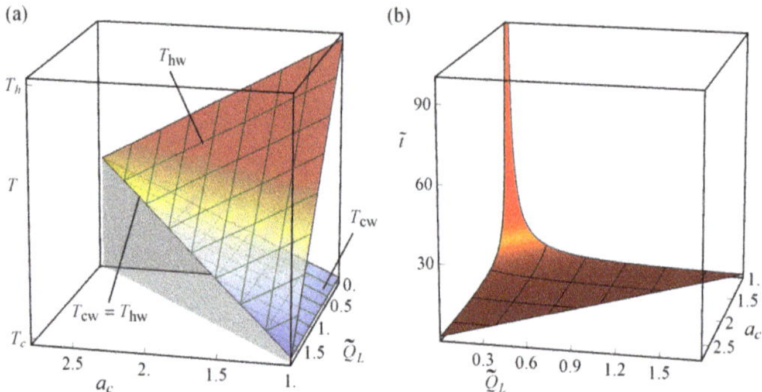

Figure 2. (a) T_{hw} and T_{cw} from Equation (9). Note how, as the heat leak increases, the possible physical combinations of T_{hw} and T_{cw} become more limited; (b) $\widetilde{t}(\widetilde{Q}_L, a_c)$ according to Equation (10). The representative values $\alpha = \frac{1}{5} = \tau$, $\widetilde{\Sigma}_c = \frac{1}{2}$ have been used, however, the displayed behavior is similar for any other combination of values.

As a heat leak appears, the reversible limit $\widetilde{t} \to \infty$ is no longer achievable. This is better reflected in Figure 2b, where we plot the total operation time $\widetilde{t}$ depending on a_c and $\widetilde{Q}_L$ (see Equation (11)). Only when $a_c \to 1$ and $\widetilde{Q}_L \to 0$ can large operation times be allowed. We can see in this figure that the existence of a heat leak imposes a maximum operational characteristic time to the HE. The total time is noticeably shorter as the heat leak increases, in agreement with the fact that, for $\widetilde{t} \leq 1$, the working regimes are dominated by dissipations. It could be said that the heat leak behaves as a causality effect in the arrow of time of the heat engine.

Notice that, in Figure 2, there is a region of prohibited combinations of $\widetilde{Q}_L$ and a_c. This has to do with the physical reality of the engine (negative power output and efficiency). In [41], the region of physical interest in the LD model under maximum power conditions was analyzed. In the Carnot-like engine, some similar considerations can be addressed as follows: in a valid endoreversible HE, the internal temperatures may vary in the range $a_h \in (1, \tau^{-1})$ and $a_c \in (1, \tau^{-1} a_h^{-1})$ in order to have $T_c \leq T_{cw} \leq T_{hw} \leq T_h$, $a_c = \tau^{-1} a_h^{-1}$ being the condition for $T_{cw} = T_{hw}$ implying null work output and efficiency. From Equation (9), it is possible to obtain two conditions on $\widetilde{Q}_L(\alpha, \widetilde{\Sigma}_c, a_c, a_h, \tau)$ (initially assumed to be ≥ 0) according to the values $a_h = 1$ and $a_h = \tau^{-1}$. For $a_h = 1$, we obtain that

$$\widetilde{Q}_L = -\frac{(a_c - 1)\left(\frac{1-\widetilde{\Sigma}_c}{\widetilde{\Sigma}_c}\right)}{\frac{1-\widetilde{\Sigma}_c}{\widetilde{\Sigma}_c} + \tau\left(\frac{1-\alpha}{\alpha}\right)} \leq 0, \tag{12}$$

whose only physical solution is $\widetilde{Q}_L = 0$. Then, as long as there is a heat leak in the device, the internal hot reservoir cannot reach equilibrium with the external hot reservoir and the reversible configuration is not achievable. On the other hand (as can be seen in Figure 2a), the largest possible heat leak (i.e., the largest dissipation in the system) has as an outcome that $T_{hw} \to T_{cw} \to T_c$, that is, $a_c \to 1$ and $a_h \to \tau^{-1}$. In that limit, Equations (9) and (11) give

$$\widetilde{Q}_{L,max} = \frac{(1-\tau)\left(\frac{1-\alpha}{\alpha}\right)}{\frac{1-\widetilde{\Sigma}_c}{\widetilde{\Sigma}_c} + \tau\left(\frac{1-\alpha}{\alpha}\right)} = \frac{\widetilde{\Sigma}_c}{\alpha \tilde{t}}, \tag{13}$$

and, since in this case all the input heat is dissipated to the cold external thermal reservoir, the HE has a null power output. In Figure 3, we depict the range of possible values that $\widetilde{Q}_L$ can take (from 0 up to $\widetilde{Q}_{L,max}$) in terms of α and $\tilde{t}$. By means of Equation (13), it is established a boundary condition for physically acceptable values of the irreversible Carnot-like HE in terms of the LD variables, which is

$$\tilde{t} = \frac{\alpha\left(1 - \widetilde{\Sigma}_c\right) + \widetilde{\Sigma}_c \tau\left(1 - \alpha\right)}{\alpha\left(1 - \alpha\right)\left(1 - \tau\right)}. \tag{14}$$

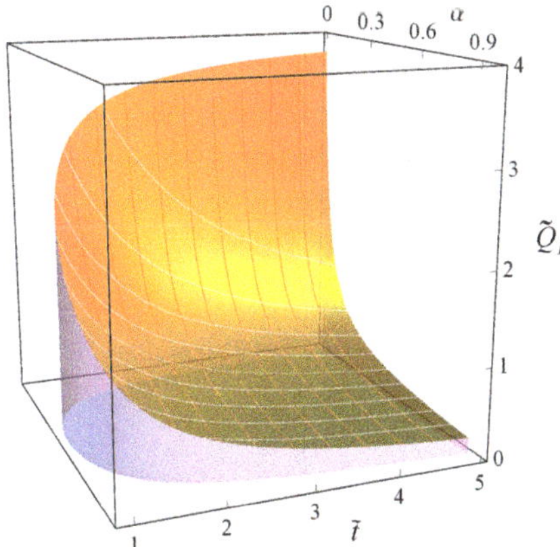

Figure 3. Possible values of $\widetilde{Q}_L$ as a function of the control parameters α and $\tilde{t}$. We used the values $\widetilde{\Sigma}_c = 0.8$ and $\tau = 0.2$.

Up to this point, we have proposed a generic correspondence between the variables of both schemes: the LD treatment, based on a specific entropy generation law, and the irreversible Carnot-like engine based on heat transfer laws. In the following, we will further analyze the connection given by Equation (11) with the focus on different heat transfer laws and the maximum power efficiencies given by the low-dissipation model.

4. Maximum-Power Regime

As is usual, the power output is given by

$$P = \frac{\eta\, Q_{\mathrm{h}}}{t_{\mathrm{c}} + t_{\mathrm{h}}}. \tag{15}$$

In [41], it was shown that, in the MP regime of an LD engine display, an open, parabolic behavior for the parametric $P - \eta$ curves when $\alpha = \alpha_{\tilde{P}_{max}}$ is fixed and for $\tilde{\Sigma}_{c} \in [0,1]$. On the other hand, by fixing the value of $\tilde{t} = \tilde{t}_{\tilde{P}_{max}}$, one obtains for the behavior of η vs. P loop-like curves (see Figure 4 in [41]). In the irreversible Carnot-like framework, open η vs. P curves are characteristic of endoreversible CA-type engines, and, when a heat leak is introduced, one obtains loop-like curves. The apparent connection between the behavior displayed by fixing $\tilde{t}$ or α in the low dissipation context with the presence of a heat leak, or the lack of it, is by no means obvious. A simple analysis of the MP regime in an irreversible Carnot-like engine in terms of the LD variables will shed some light on this issue and will also provide us a better understanding of how good the correspondence is between both schemes.

4.1. Low Dissipation Heat Engine

In terms of the characteristic variables, the input and output heat are

$$\dot{\tilde{Q}}_{\mathrm{h}} \equiv \frac{\tilde{Q}_{\mathrm{h}}}{\tilde{t}} = \frac{Q_{\mathrm{h}}}{T_{\mathrm{c}}\Delta S}\frac{\Sigma_{\mathrm{T}}}{t\,\Delta S} = \left(1 - \frac{1 - \tilde{\Sigma}_{\mathrm{c}}}{(1-\alpha)\tilde{t}}\right)\frac{1}{\tau\tilde{t}}, \tag{16}$$

$$\dot{\tilde{Q}}_{\mathrm{c}} \equiv \frac{\tilde{Q}_{\mathrm{c}}}{\tilde{t}} = \frac{Q_{\mathrm{c}}}{T_{\mathrm{c}}\Delta S}\frac{\Sigma_{\mathrm{T}}}{t\,\Delta S} = \left(1 - \frac{\tilde{\Sigma}_{\mathrm{c}}}{\alpha\tilde{t}}\right)\frac{1}{\tilde{t}}, \tag{17}$$

giving a power output and efficiency as follows:

$$\tilde{P} \equiv -\frac{\tilde{W}}{\tilde{t}} = -\frac{W}{T_{\mathrm{c}}\Delta S}\frac{\Sigma_{\mathrm{T}}}{t\,\Delta S} = \left[\frac{1}{\tau} - 1 - \frac{1}{\tau}\left(\frac{1 - \tilde{\Sigma}_{\mathrm{c}}}{(1-\alpha)\tilde{t}}\right) - \frac{\tilde{\Sigma}_{\mathrm{c}}}{\alpha\tilde{t}}\right]\frac{1}{\tilde{t}}, \tag{18}$$

$$\tilde{\eta} \equiv \frac{\tilde{P}}{\dot{\tilde{Q}}_{\mathrm{h}}} = -\frac{W}{Q_{\mathrm{h}}} = \eta = \frac{1 - \tau - \frac{1-\tilde{\Sigma}_{\mathrm{c}}}{(1-\alpha)\tilde{t}} - \frac{\tau\tilde{\Sigma}_{\mathrm{c}}}{\alpha t}}{1 - \frac{1-\tilde{\Sigma}_{\mathrm{c}}}{(1-\alpha)\tilde{t}}}. \tag{19}$$

The optimization of $\tilde{P}(\tilde{t}, \alpha; \tilde{\Sigma}_{\mathrm{c}}, \tau)$ is accomplished through the partial contact time α and the total time $\tilde{t}$, whose values are

$$\alpha_{\tilde{P}_{max}}\left(\tilde{\Sigma}_{\mathrm{c}}, \tau\right) = \frac{1}{1 + \sqrt{\frac{1-\tilde{\Sigma}_{\mathrm{c}}}{\tau\tilde{\Sigma}_{\mathrm{c}}}}}, \tag{20}$$

$$\tilde{t}_{\tilde{P}_{max}}\left(\tilde{\Sigma}_{\mathrm{c}}, \tau\right) = \frac{2}{1-\tau}\left(\sqrt{\tau\tilde{\Sigma}_{\mathrm{c}}} + \sqrt{1-\tilde{\Sigma}_{\mathrm{c}}}\right)^{2}, \tag{21}$$

with an MP efficiency given by

$$\eta_{\tilde{P}_{max}}\left(\tilde{\Sigma}_c, \tau\right) = \frac{(1-\tau)\left[1 + \sqrt{\frac{\tau\tilde{\Sigma}_c}{1-\tilde{\Sigma}_c}}\right]}{\left[1 + \sqrt{\frac{\tau\tilde{\Sigma}_c}{1-\tilde{\Sigma}_c}}\right]^2 + \tau\left(1 - \frac{\tilde{\Sigma}_c}{1-\tilde{\Sigma}_c}\right)}. \tag{22}$$

One of the most relevant features of this model is the capability of obtaining upper and lower bounds of the MP efficiencies without any information regarding the heat fluxes nature. These limits are

$$\eta_{\tilde{P}_{max}}^{-} = \frac{\eta_C}{2} \leq \eta_{\tilde{P}_{max}}(\tau, \tilde{\Sigma}_c) \leq \frac{2 - \eta_C}{\eta_C} = \eta_{\tilde{P}_{max}}^{+}, \tag{23}$$

corresponding to $\tilde{\Sigma}_c = 1$ and $\tilde{\Sigma}_c = 0$ for the lower and upper bounds, respectively. For the symmetric dissipation case, $\tilde{\Sigma}_c = 1/2(= \tilde{\Sigma}_h)$, the well known CA-efficiency $\eta_{\tilde{P}_{max}}^{sym} = 1 - \sqrt{\tau} = \eta_{CA}$ is recovered.

4.2. Carnot-Like Model without Heat Leak (Endoreversible Model)

Now, let us consider a family of heat transfer laws depending on the power of the temperature to model the heat fluxes Q_h and Q_c (see Figure 1b) as follows:

$$Q_h = T_h^k \sigma_h \left(1 - a_h^{-k}\right) \quad t_h \geq 0, \tag{24}$$

$$Q_c = T_c^k \sigma_c \left(a_c^k - 1\right) \quad t_c \geq 0, \tag{25}$$

where $k \neq 0$ is a real number, σ_h and σ_c are the conductances in each process and t_h and t_c are the times at which the isothermal processes are completed. The adiabatic processes are considered as instantaneous, a common assumption in the two models. According to Equation (15), power output is a function depending on the variables a_c, a_h and the ratio of contact times; k, τ and σ_{hc} are not optimization variables for this model. The endoreversible hypothesis $\Delta S_{T_{hw}} = -\Delta S_{T_{cw}}$ gives the following constriction upon the contact times ratio

$$\frac{t_c}{t_h} = \sigma_{hc} a_c a_h \tau^{1-k} \left(\frac{1 - a_h^{-k}}{a_c^k - 1}\right), \tag{26}$$

where $\sigma_{hc} \equiv \sigma_h/\sigma_c$. Since there is no heat leak, the efficiency of the internal Carnot cycle is the same as the efficiency of the engine, then $a_c a_h \tau = 1 - \eta$, and the dependence of a_h is substituted by η. Then, in terms of α, Equation (26) is

$$\frac{\alpha}{1 - \alpha} = \sigma_{hc} \tau^k (1 - \eta) \left(\frac{1 - \frac{a_c^k \tau^k}{(1-\eta)^k}}{a_c^k - 1}\right). \tag{27}$$

The optimization of power output $P\left(a_c, \eta; \sigma_{hc}, \tau, T_h, k\right)$ in this case is achieved through a_c and η. The maximum power is obtained by solving $\left(\frac{\partial P}{\partial a_c}\right)_\eta = 0$ for a_c and $\left(\frac{\partial P}{\partial \eta}\right)_{a_c} = 0$ for η. From the first condition, we obtain P^*, which is

$$P^*\left(\eta; \sigma_{hc}, \tau, T_h, k\right) = \sigma_h T_h^k \left(\frac{\eta}{1 - \eta}\right) \frac{(1 - \eta)^k - \tau^k}{\left(\sqrt{\sigma_{hc}} + (1 - \eta)^{\frac{k-1}{2}}\right)^2}. \tag{28}$$

This function has a unique maximum corresponding to $\eta_{P_{max}}$, which is the solution to the following equation

$$\sqrt{\sigma_{hc}}\,(1-\eta)\left[\tau^k - (1-\eta)^k\,(1-k\eta)\right] + (1-\eta)^{\frac{k+1}{2}}\left[(1-(1-k)\,\eta)\,\tau^k - (1-\eta)^{k+1}\right] = 0, \qquad (29)$$

and depends on the values σ_{hc}, τ and the exponent of the heat transfer law k as showed in [9]. In Figure 4a, $\eta_{P_{max}}$ is depicted for the limiting cases $\sigma_{hc} \to \{0, \infty\}$. All of the possible values of $\eta_{P_{max}}$ for different σ_{hc}s are located between these two curves. It is well-known that, for the Newtonian heat transfer law ($k = 1$), $\eta_{P_{max}} = \eta_{CA}$ is independent of the σ_{hc} value. As the heat transfer law departs from the Newtonian case, the upper and lower bounds cover a wider range of efficiencies. Then, the limits appearing in Equation (23) are fulfilled for a limited region of k values in the Carnot-like model. From Figure 4a, it is possible to see that the results stemming from an endoreversible engine are accessible from an LD landmark only in the region $k \in (-1, 2.5]$ (for other values of k, there are efficiencies outside the range given by Equation (23)). By equaling these efficiencies with the LD one (see Equation (22)) and solving for $\widetilde{\Sigma}_c$, we obtain those values that reproduce the endoreversible efficiencies. This is depicted in Figure 4b. Notice also that not all $\widetilde{\Sigma}_c$ symmetries are allowed for every $k \in (-1, 2.5]$. For example, with a heat transfer law with exponent $k = -1$, all the possible values of the efficiency can be obtained if the parameter $\widetilde{\Sigma}_c$ varies from 0 to 1, that is, all symmetries are allowed. Meanwhile, for $k = 1$, only the symmetric case $\widetilde{\Sigma}_c = 1/2$ is allowed, reproducing the CA efficiency. For k outside $(-1, 2.5]$, there are efficiencies above and below the limits in Equation (23) with no Σ_c values that might reproduce those efficiencies, thus limiting the heat transfer laws physically consistent with predictions of the LD model.

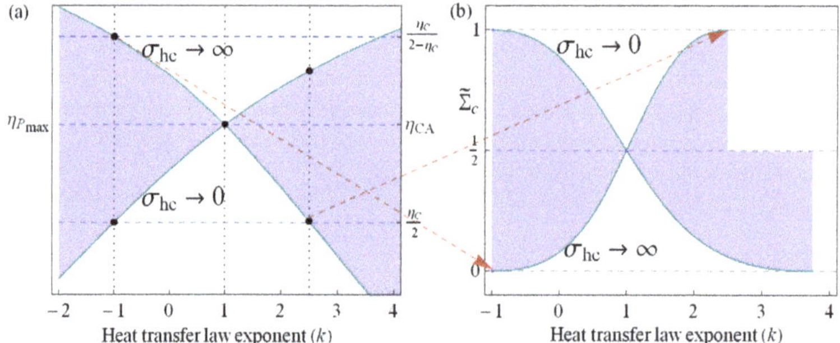

Figure 4. (**a**) upper and lower bounds of the MP efficiency in terms of the exponent of the heat transfer law k of the Carnot-like heat engine; (**b**) the $\widetilde{\Sigma}_c$ values that reproduce the upper and lower bounds of the endoreversible engine.

Inside the region where the LD model is able to reproduce the asymmetric limiting cases ($\sigma_{hc} \to \{0, \infty\}$), the correspondence between the two formalisms has not an exact fitting. In order to show this, we will address the symmetric dissipation case.

As can be seen from Figure 4, in the endoreversible CA-type HE, for every k, there is one σ_{hc} that reproduces the CA efficiency. On the other hand, in the LD model, the symmetric dissipation is attached to η_{CA}. If we use the α and $\tilde{t}$ values of MP of the LD model and calculate the values of a_c and a_h associated with them (instead of calculating them according to Equations (28) and (29)), we can see whether they allow us to recover the correct value of σ_{hc} that in the endoreversible model gives the CA efficiency or does not. That is, for $\widetilde{\Sigma}_c = 1/2$, Equations (20) and (21) reduce to

$$\alpha_{\widetilde{P}_{max}}^{sym} = \frac{\sqrt{\tau}}{1 + \sqrt{\tau}}, \tag{30}$$

$$\tilde{t}_{\widetilde{P}_{max}}^{sym} = \frac{1 + \sqrt{\tau}}{1 - \sqrt{\tau}}. \tag{31}$$

From Equation (11), we obtain a_c and with the condition $\eta = \eta_{CA}$, with $T_{cw}/T_{hw} = a_c a_h \tau = 1 - \eta = \sqrt{\tau}$ we calculate a_h, thus

$$a_{c,\widetilde{P}_{max}}^{sym} = \frac{1 + \sqrt{\tau}}{2\sqrt{\tau}}, \tag{32}$$

$$a_{h,\widetilde{P}_{max}}^{sym} = \frac{2}{1 + \sqrt{\tau}}. \tag{33}$$

By using Equation (30), the ratio of contact times results in $\frac{t_c}{t_h} = \frac{\alpha}{1-\alpha} = \sqrt{\tau}$, and, by using the endoreversible hypothesis (Equation (26)), it is possible to obtain the value of σ_{hc} that would produce the CA efficiency, being

$$\sigma_{hc,\widetilde{P}_{max}}^{sym} = \frac{\left(1 + \sqrt{\tau}\right)^k \tau^{\frac{k}{2}} - 2^k \iota^k}{2^k - \left(1 + \sqrt{\tau}\right)^k}, \tag{34}$$

which for $k = 1$ gives $\sigma_{hc,\widetilde{P}_{max}}^{sym} = \sqrt{\tau}$ and for $k = -1$ gives $\sigma_{hc,\widetilde{P}_{max}}^{sym} = 1/\tau$. Nevertheless, by substituting Equation (34) into Equation (29), the MP efficiency is not exactly the CA one, as can be seen in Figure 5. Showing that the correspondence between both models is a good approximation only in the range $k \in [-1, 1]$, and is exact only for $k = -1$ and $k = 1$.

Another incompatibility of the two approaches comes up in the Newtonian heat exchange ($k = 1$): meanwhile, the Carnot-like scheme η_{CA} is independent of any value of σ_{hc}, and, in terms of the LD model, η_{CA} is strictly attached to a symmetric dissipation $\widetilde{\Sigma}_c = 1/2 (= \widetilde{\Sigma}_h)$. Then, the only law that has an exact correspondence for all values of $\widetilde{\Sigma}_c$ and σ_{hc} is the law $k = -1$.

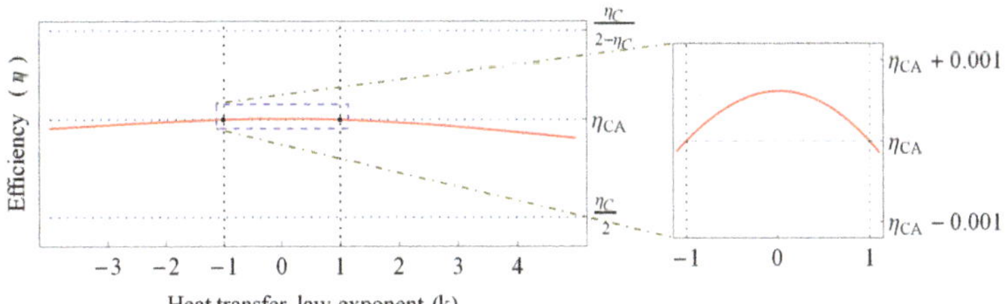

Figure 5. Maximum-power efficiency for the symmetric case $\widetilde{\Sigma}_c = 1/2$, assuming the LD condition that $\frac{t_c}{t_h} = \sqrt{\tau}$ and using the resulting σ_{hc} value that fulfills the endoreversible hypothesis. Notice that the matching with the CA efficiency is approximate for the interval $k \in [-1, 1]$ and is exact for $k = \{-1, 1\}$, as can be seen in the zoom of this region on the right side of the figure.

4.3. Carnot-Like Model with Heat Leak

Now, let us consider a heat leak of the same kind of the heat fluxes Q_c and Q_h, that is,

$$Q_L = T_h^k \sigma_L \left(1 - \tau^k\right) (t_h + t_c) \geq 0, \tag{35}$$

then, the characteristic heat leak is

$$\widetilde{Q}_{\mathrm{L}} = \frac{Q_{\mathrm{L}}}{Q_{\mathrm{c}}} = a_{\mathrm{c}} \frac{T_{\mathrm{h}}^k \sigma_{\mathrm{L}} \left(1 - \tau^k\right)(t_{\mathrm{h}} + t_{\mathrm{c}})}{T_{\mathrm{c}}^k \sigma_{\mathrm{c}} \left(a_{\mathrm{c}}^k - 1\right) t_{\mathrm{c}}} = \frac{a_{\mathrm{c}} \sigma_{\mathrm{Lc}} \left(1 - \tau^k\right)}{\alpha \, \tau^k \left(a_{\mathrm{c}}^k - 1\right)}. \tag{36}$$

The power output of the engine is the same than in the endoreversible case; however, a difference with the previous subsection arises, and now the efficiency is given by the following expression:

$$\eta = \frac{Q_{\mathrm{h}} + Q_{\mathrm{L}} - Q_{\mathrm{c}} - Q_{\mathrm{L}}}{Q_{\mathrm{h}} + Q_{\mathrm{L}}} = 1 - \frac{Q_{\mathrm{c}} + Q_{\mathrm{L}}}{Q_{\mathrm{h}} + Q_{\mathrm{L}}} = 1 - \frac{1 + \frac{Q_{\mathrm{L}}}{Q_{\mathrm{c}}}}{\frac{Q_{\mathrm{h}}}{Q_{\mathrm{c}}} + \frac{Q_{\mathrm{L}}}{Q_{\mathrm{c}}}} = 1 - \frac{1 + \widetilde{Q}_{\mathrm{L}}}{\widetilde{Q}_{\mathrm{h}} + \widetilde{Q}_{\mathrm{L}}}, \tag{37}$$

where we have used the fact that $T_{\mathrm{c}}\Delta S = Q_{\mathrm{c}}$ to introduce the characteristic heats in the last expression.

From Equation (37), it can be derived that, if the heat leak increases, the efficiency diminishes. In Figure 6, we can observe how the upper and lower bounds of the efficiency appearing in Figure 4 are affected by the introduction of a constant heat leak. Now, by using Equation (37) and the fact that in the endoreversible case $\eta = 1 - \widetilde{Q}_{\mathrm{h}}^{-1}$, we have plotted the $\widetilde{Q}_{\mathrm{L}}$ values that leads the efficiency η_{CA} for $k = 1$ (of the endoreversible case) down to the value $\eta_C/2$ (see point A in Figure 6), which is $\widetilde{Q}_{\mathrm{L}} = \frac{1 - \sqrt{\tau}}{\sqrt{\tau}(1 + \sqrt{\tau})}$; and the value $\widetilde{Q}_{\mathrm{L}} = \frac{1 - \tau}{2\tau}$ that lead to an efficiency $\eta_C/(2 - \eta_C)$ for $k = -1$ (of the endoreversible case) to $\eta_C/2$ (see point B in Figure 6).

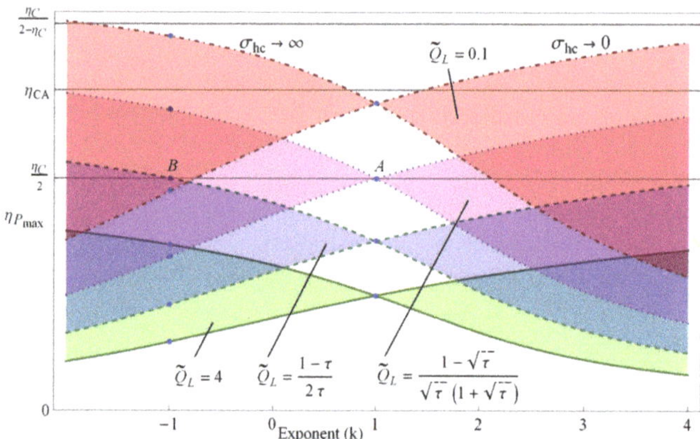

Figure 6. Influence of the heat leak over the optimized efficiencies appearing in Figure 4. See the text for explanation.

Notice in Figure 6 that, for $\widetilde{Q}_{\mathrm{L}} = 0.1$, there is a region around $k = 1$ where η_{CA} is outside the shaded region of maximum power efficiencies. Then, for these values of k and $\widetilde{Q}_{\mathrm{L}}$, the symmetric dissipation case, always attached to the η_{CA} efficiency, is out of reach. Additionally, there is not a heat transfer law that fulfills both upper and lower bounds for maximum power efficiency given by the LD model as occurred for $k = -1$ in the endoreversible case. The additional degree of freedom caused by the appearance of the heat leak makes more complex the analysis of the validity of the correspondence between both models, which, in general, should be handled numerically.

5. Conclusions

We analyzed Carnot-like heat engines (dependent on heat transfer laws) and the LD models (dependent on a specific entropy generation law) and studied how the variables of each one are connected. We were able to provide an interpretation of the heat transfer laws, including the heat leak, in terms of the bounds for the efficiency at MP provided by the LD-model, which, in turn, are dependent on the relative symmetries of the dissipations constants and the partial contact times.

By comparing the entropy production of the low dissipation model and the Carnot-like model, we proposed a connection between the variables that describe each model. We show that, for an HE, the region of physical interest is independent from the operation regime, being equivalent to that for a maximum power LD-HE. That is, $\eta\left(\tilde{t}_{\tilde{P}_{max}}, \alpha, \tilde{\Sigma}_{c}, \tau\right) \geq 0$ defines the acceptable α values, and $\eta\left(\tilde{t}, \alpha_{\tilde{P}_{max}}, \tilde{\Sigma}_{c}, \tau\right) \geq 0$ those of $\tilde{t}$ (gray shaded areas in Figure 7a,b, respectively). These considerations on the LD model are recovered from physical considerations on the Carnot-like HE. Thus, the difference in the performance of the HEs is not due to the physical configurations of the system, but it comes from the approximations that these models rely on: one over the entropy and the other over the heat fluxes.

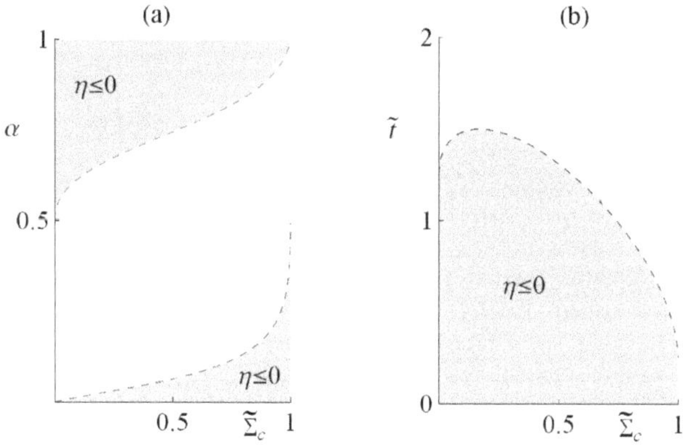

Figure 7. (**a**) Physically well behaved region of the α–$\tilde{\Sigma}_c$ variables. The shaded areas come from the LD model and the dashed curves come from the Carnot-like model; (**b**) The same for the $\tilde{t}$–$\tilde{\Sigma}_c$ variables. Notice the agreement in both models. In these plots, we use $\tau = 0.2$.

We show that the heat leak disappears when fixing the partial contact time in the LD engine, leading to typical endoreversible open parabolic power vs. efficiency curves. However, the connection between the endoreversible case and the LD model is exact only for heat transfer laws with exponents $k = 1$ and $k = -1$, and a good approximation in the region $k \in (-1, 1)$. On the other hand, the presence of a heat leak fixes the total operation time and the partial contact time is not constrained, thus, allowing the heat leak to act as an additional degree of freedom (the same efficiency is achieved with different combinations of partial contact time ratios and heat leaks). The reversible limit is not accessible in this case, a maximum operation time is established, and when the heat leak dissipation effects are important, the efficiency may be zero, which is the origin of the loop behavior of the irreversible $\tilde{P}$ vs. η curves. The connection in the case with heat leak is more complex and its validity depends on the value of $\tilde{Q}_L$.

Acknowledgments: Julian Gonzalez-Ayala acknowledges CONACYT-MÉXICO; José Miguel M. Roco, Alejandro Medina and Antonio Calvo-Hernández acknowledge the Ministerio de Economía y Competitividad (MINECO) of Spain for financial support under Grant ENE2013-40644-R.

Author Contributions: José Miguel M. Roco and Antonio Calvo Hernández conceived and made some simulations of the thermodynamic models. Julian Gonzalez-Ayala and Alejandro Medina. complemented calculations, simulations and the elaboration of graphics. Julian Gonzalez-Ayala and Antonio Calvo Hernández wrote the paper. All authors have read and approved the final manuscript. All authors have read and approved the final manuscript.

Conflicts of Interest: The authors declare no conflict of interest.

References

1. Curzon, F.L.; Ahlborn, B. Efficiency of a Carnot Engine at Maximum Power Output. *Am. J. Phys.* **1975**, *43*, 22–24.
2. Vaudrey, A.; Lanzetta, F.; Feidt, M. H. B. Reitlinger and the origins of the Efficiency at Maximum Power formula for Heat Engines. *J. Non-Equilib. Thermodyn.* **2014**, *39*, 199–203.
3. Calvo Hernández, A.; Roco, J.M.M.; Medina, A.; Velasco, S.; Guzmán-Vargas, L. The maximum power efficiency 1-root tau: Research, education, and bibliometric relevance. *Eur. Phys. J. Spec. Topics* **2015**, *224*, 809–821.
4. Arias-Hernandez, L.A.; Angulo-Brown, F.; Paez-Hernandez, R.T. First-order irreversible thermodynamic approach to a simple energy converter. *Phys. Rev. E* **2008**, *77*, 011123.
5. Izumida, Y.; Okuda, K. Efficiency at maximum power of minimally nonlinear irreversible heat engines. *Europhys. Lett.* **2012**, *97*, 10004.
6. Apertet, Y.; Ouerdane, H.; Goupil, C.; Lecoeur, P. Irreversibilities and efficiency at maximum power of heat engines: The illustrative case of a thermoelectric generator. *Phys. Rev. E* **2012**, *85*, 031116.
7. Wang, Y.; Tu, Z.C. Efficiency at maximum power output of linear irreversible Carnot-like heat engines. *Phys. Rev. E* **2012**, *85*, 011127.
8. Apertet, Y.; Ouerdane, H.; Goupil, C.; Lecoeur, P. Efficiency at maximum power of thermally coupled heat engines. *Phys. Rev. E* **2012**, *85*, 041144.
9. Gonzalez-Ayala, J.; Arias-Hernandez, L.A.; Angulo-Brown, F. Connection between maximum-work and maximum-power thermal cycles. *Phys. Rev. E* **2013**, *88*, 052142.
10. Sheng, S.; Tu, Z.C. Universality of energy conversion efficiency for optimal tight-coupling heat engines and refrigerators. *J. Phys. A Math. Theor.* **2013**, *46*, 402001.
11. Aneja, P.; Katyayan, H.; Johal, R.S. Optimal engine performance using inference for non-identical finite source and sink. *Mod. Phys. Lett. B* **2015**, *29*, 1550217.
12. Sheng, S.Q.; Tu, Z.C. Constitutive relation for nonlinear response and universality of efficiency at maximum power for tight-coupling heat engines. *Phys. Rev. E* **2015**, *91*, 022136.
13. Cleuren, B.; Rutten, B.; van den Broeck, C. Universality of efficiency at maximum power: Macroscopic manifestation of microscopic constraints. *Eur. Phys. J. Spec. Topics* **2015**, *224*, 879–889.
14. Izumida, Y.; Okuda, K. Linear irreversible heat engines based on the local equilibrium assumptions. *New J. Phys.* **2015**, *17*, 085011.
15. Long, R.; Liu, W. Efficiency and its bounds of minimally nonlinear irreversible heat engines at arbitrary power. *Phys. Rev. E* **2016**, *94*, 052114.
16. Wang, Y. Optimizing work output for finite-sized heat reservoirs: Beyond linear response. *Phys. Rev. E* **2016**, *93*, 012120.
17. Schmiedl, T.; Seifert, U. Optimal Finite-Time Processes in Stochastic Thermodynamics. *Phys. Rev. Lett.* **2007**, *98*, 108301.
18. Tu, Z.C. Efficiency at maximum power of Feynman's ratchet as a heat engine. *J. Phys. A Math. Theor.* **2008**, *41*, 312003.
19. Schmiedl, T.; Seifert, U. Efficiency at maximum power: An analytically solvable model for stochastic heat engines. *Europhys. Lett.* **2008**, *81*, 20003.
20. Schmiedl, T.; Seifert, U. Efficiency of molecular motors at maximum power. *Europhys. Lett.* **2008**, *83*, 30005.
21. Esposito, M.; Lindenberg, K.; van den Broeck, C. Universality of Efficiency at Maximum Power. *Phys. Rev. Lett.* **2009**, *102*, 130602.
22. Esposito, M.; Lindenberg, K.; van den Broeck, C. Thermoelectric efficiency at maximum power in a quantum dot. *Europhys. Lett.* **2009**, *85*, 60010.
23. Golubeva, N.; Imparato, A. Efficiency at Maximum Power of Interacting Molecular Machines. *Phys. Rev. Lett.* **2012**, *109*, 190602.
24. Wang, R.; Wang, J.; He, J.; Ma, Y. Efficiency at maximum power of a heat engine working with a two-level atomic system. *Phys. Rev. E* **2013**, *87*, 042119.
25. Uzdin, R.; Kosloff, R. Universal features in the efficiency at maximum work of hot quantum Otto engines. *Europhys. Lett.* **2014**, *108*, 40001.

26. Curto-Risso, P.L.; Medina, A.; Calvo Hernández, A. Theoretical and simulated models for an irreversible Otto cycle. *J. Appl. Phys.* **2008**, *104*, 094911.

27. Curto-Risso, P.L.; Medina, A.; Calvo Hernández, A. Optimizing the operation of a spark ignition engine: Simulation and theoretical tools. *J. Appl. Phys.* **2009**, *105*, 094904.

28. Correa, L.A.; Palao, J.P.; Alonso, D. Internal dissipation and heat leaks in quantum thermodynamic cycles. *Phys. Rev. E* **2015**, *92*, 032136.

29. Sánchez-Salas, N.; López-Palacios, L.; Velasco, S.; Calvo Hernández, A. Optimization criteria, bounds, and efficiencies of heat engines. *Phys. Rev. E* **2010**, *82*, 051101.

30. Zhang, Y.; Huang, C.; Lin, G.; Chen, J. Universality of efficiency at unified trade-off optimization. *Phys. Rev. E* **2016**, *93*, 032152.

31. Iyyappan, I.; Ponmurugan, M. Thermoelectric energy converters under a trade-off figure of merit with broken time-reversal symmetry. *arXiv* **2016**, arXiv:1604.07551.

32. Angulo-Brown, F. An ecological optimization criterion for finite-time heat engines. *J. Appl. Phys.* **1991**, *69*, 7465–7469.

33. Arias-Hernández, L.A.; Angulo-Brown, F. A general property of endoreversible thermal engines. *J. Appl. Phys.* **1997**, *81*, 2973–2979.

34. Long, R.; Li, B.; Liu, Z.; Liu, W. Ecological analysis of a thermally regenerative electrochemical cycle. *Energy* **2016**, *107*, 95–102.

35. Esposito, M.; Kawai, R.; Lindenberg, K.; van den Broeck, C. Efficiency at Maximum Power of Low-Dissipation Carnot Engines. *Phys. Rev. Lett.* **2010**, *105*, 150603.

36. De Tomás, C.; Roco, J.M.M.; Calvo Hernández, A.; Wang, Y.; Tu, Z.C. Low-dissipation heat devices: Unified trade-off optimization and bounds. *Phys. Rev. E* **2013**, *87*, 012105.

37. Holubec, V.; Ryabov, A. Efficiency at and near maximum power of low-dissipation heat engines. *Phys. Rev. E* **2015**, *92*, 052125.

38. Holubec, V.; Ryabov, A. Erratum: Efficiency at and near maximum power of low-dissipation heat engines. *Phys. Rev. E* **2015**, *93*, 059904.

39. Holubec, V.; Ryabov, A. Maximum effciency of low-dissipation heat engines at arbitrary power. *J. Stat. Mech.* **2016**, *2016*, 073204.

40. Calvo Hernández, A.; Medina, A.; Roco, J.M.M. Time, entropy generation, and optimization in low-dissipation heat devices. *New J. Phys.* **2015**, *17*, 075011.

41. Gonzalez-Ayala, J.; Calvo Hernández, A.; Roco, J.M.M. Irreversible and endoreversible behaviors of the LD-model for heat devices: The role of the time constraints and symmetries on the performance at maximum χ figure of merit. *J. Stat. Mech.* **2016**, *2016*, 073202.

42. Gonzalez-Ayala, J.; Calvo Hernández, A.; Roco, J.M.M. From maximum power to a trade-off optimization of low-dissipation heat engines: Influence of control parameters and the role of entropy generation. *Phys. Rev. E* **2017**, *95*, 022131.

43. Hu, Y.; Wu, F.; Ma, Y.; He, J.; Wang, J.; Calvo Hernández, A.; Roco, J.M.M. Coefficient of performance for a low-dissipation Carnot-like refrigerator with nonadiabatic dissipation. *Phys. Rev. E* **2013**, *88*, 062115.

Perspective

The History and Perspectives of Efficiency at Maximum Power of the Carnot Engine

Michel Feidt

Laboratoire d'Énergétique et de Mécanique Théorique et Appliquée, UMR 7563, Université de Lorraine, 54518 Vandoeuvre-lès-Nancy, France; Michel.Feidt@univ-lorraine.fr; Tel.: +33-383-595-554

Received: 30 March 2017; Accepted: 14 July 2017; Published: 19 July 2017

Abstract: Finite Time Thermodynamics is generally associated with the Curzon–Ahlborn approach to the Carnot cycle. Recently, previous publications on the subject were discovered, which prove that the history of Finite Time Thermodynamics started more than sixty years before even the work of Chambadal and Novikov (1957). The paper proposes a careful examination of the similarities and differences between these pioneering works and the consequences they had on the works that followed. The modelling of the Carnot engine was carried out in three steps, namely (1) modelling with time durations of the isothermal processes, as done by Curzon and Ahlborn; (2) modelling at a steady-state operation regime for which the time does not appear explicitly; and (3) modelling of transient conditions which requires the time to appear explicitly. Whatever the method of modelling used, the subsequent optimization appears to be related to specific physical dimensions. The main goal of the methodology is to choose the objective function, which here is the power, and to define the associated constraints. We propose a specific approach, focusing on the main functions that respond to engineering requirements. The study of the Carnot engine illustrates the synthesis carried out and proves that the primary interest for an engineer is mainly connected to what we called Finite (physical) Dimensions Optimal Thermodynamics, including time in the case of transient modelling.

Keywords: Carnot engine; modelling with time durations; steady-state modelling; transient conditions; converter irreversibility; sequential optimization; Finite physical Dimensions Optimal Thermodynamics

---

1. Introduction

The development of Finite Time Thermodynamics is generally associated with Curzon and Ahlborn's paper [1], and concerns the efficiency of a Carnot engine at Maximum Power output. This is the core subject of the present paper.

This problem is of fundamental importance because the efficiency of energy use is one of the most important goals all over the world for the future of our civilization.

Carnot was one of the first to introduce the concept of efficiency related to energy conversion from thermal energy to mechanical energy. However, this work was in the framework of Equilibrium Thermodynamics and accordingly, with zero power due to quasi static transformations implying infinite time duration as well.

Nevertheless, the First Law efficiency corresponds to a well-known upper bound, namely:

$$\eta_I = 1 - \frac{T_{CS}}{T_{HS}} \tag{1}$$

The originality of Curzon and Ahlborn's paper was that it related efficiency to the maximum power output $\dot{W}$ of an endoreversible Carnot engine, as will be detailed and discussed below.

Using the same hypothesis as Carnot regarding infinite Hot Source at T_{HS}, and infinite Cold Sink at T_{CS}, they obtained what is called the nice radical:

$$\eta_I \left(MAX \ \dot{W} \right) = 1 - \sqrt{\frac{T_{CS}}{T_{HS}}} \tag{2}$$

However, as we indicated in a publication in the 1980s, the nice radical was present in the publications of Chambadal [2] and also Novikov [3], but for slightly different models, as will be specified in the following sections. Years ago, we also discovered that Reitlinger had been involved with the origins of efficiency at maximum power [4], as stated in a paper published in Liège (Belgium) in 1929. Moreover, a presentation [5] at a conference held in Bucharest in June 2016 revealed that in Moutier's book [6] (p. 62) published in 1872, he introduced a nice radical that was named an "economical coefficient" (see Figure 1).

(62)

en jetant les yeux sur l'une des *fig.* 10 ou 11, on voit immédiatement que le travail effectué par la machine tend à disparaître, en même temps que le coefficient économique de la machine se rapproche de la valeur maximum.

La question industrielle ne consiste pas seulement à augmenter le coefficient économique, mais à obtenir le plus de travail possible d'une machine qui fonctionne entre des limites déterminées de température, qui sont ici T_1 et T_2.

Pour atteindre ce but, il faut disposer de T de manière à rendre $q - q'$ maximum, ou, ce qui est la même chose, de manière à rendre minimum

$$T + \frac{T_1 T_2}{T}.$$

Or, le produit $T \times \frac{T_1 T_2}{T} = T_1 T_2$ étant constant, il faut que les deux facteurs T et $\frac{T_1 T_2}{T}$ soient égaux,

$$T = \frac{T_1 T_2}{T},$$

$$T = \sqrt{T_1 T_2}.$$

Dans ce cas remarquable,

$$T' = \frac{T_1 T_2}{T} = T;$$

ainsi, le travail effectué par l'une des machines précédentes est maximum lorsque les températures intermédiaires T et T' ont pour valeur commune une moyenne proportionnelle entre les limites de température T_1 et T_2.

Le coefficient économique dans ce cas est

$$1 - \sqrt{\frac{T_2}{T_1}}.$$

Figure 1. Extract from Moutier's book [6] (p. 62).

The optimum considered in Moutier's approach is the maximum of the mechanical work obtained from the available heat.

In a more recent thesis published in Paris [7], we found that other scientific works from the past [8,9] discussed the same approach. Serrier's book [9] (published in 1888) discusses the calculation of a maximum work per cycle between a maximum temperature, T_1, and a minimum one, T_3, (author notation), such that the work W was expressed as:

$$MAX(W) = C \cdot \left(\sqrt{T_1} - \sqrt{T_3} \right)^2 \tag{3}$$

with a constant coefficient for the transfer of heat.

The corresponding "economical coefficient" (in fact η_I, first law efficiency) corresponds to the nice radical. Furthermore, the ratio of this coefficient to the Carnot efficiency (called η_{II}, second law efficiency) was given. The author adds here that the maximum work differs if a given available heat is imposed.

Furthermore, since the 1990s, we have collaborated with S. Petrescu's team at the University Politehnica of Bucharest [10] on what is named Finite Speed Thermodynamics.

It is remarkable to see that the problem of conversion (today called valorization) of heat to mechanical energy had already been effectively developed more than a century and a half ago, with the main objective being maximum work or power allied with efficiency [11].

More recent achievements include the extension of scientific and technical studies to more numerous potential applications (namely other engines) but also, from a more fundamental point of view, the development of new upper bounds in term of efficiency at maximum power.

The present paper proposes to analyze (Section 2) the similarities and differences found in the models developed during the first century (until Curzon and Ahlborn's seminal paper). Consequently, Section 3 will be devoted to steady-state modelling focused on the Carnot engine. This implies important conclusions for thermo-mechanical engines.

Section 4 is concerned with transient modelling, a new branch of Finite Time Thermodynamics that depends explicitly on time but in a specific form. This will be highlighted in the section.

In Section 5, results are discussed and summarized showing the evolution of knowledge about maximum power and efficiency since Carnot's pioneering work. Some recommendations and remarks are given regarding links with practical aspects. Finally, the future perspectives of what we have called Finite Dimension Optimal Thermodynamics (FDOT) are proposed. It could be said that FDOT corresponds to a unification of various branches of new tendencies in Thermodynamics which are Finite Time Thermodynamics (FTT), Finite Speed Thermodynamics (FST), and also Finite Size Thermodynamics. New results regarding efficiency at maximum power are proposed. When consistent, the optimal physical dimensions allocation is given. It corresponds to G_i optimal distribution, G_i being the physical dimension concerned.

2. Finite Time Thermodynamics.

2.1. Equilibrium Thermodynamics Limit

The work of Carnot was relative to the mechanical work W, and the corresponding First Law efficiency η_I defined as the ratio of useful mechanical energy, W, to the heat expense, Q_H:

$$\eta_I = \frac{W}{Q_H} \tag{4}$$

This definition is general, and could represent any First Law efficiency, whatever the model and associated hypothesis.

We shall first consider the Equilibrium Thermodynamics limit. It is well known that for the Carnot cycle representing an engine with perfect thermal contacts at the isothermal source (T_{HS}) and sink

(T_{CS}), the efficiency is given by Equation (1). This expression is valid for a reversible engine considered as an adiabatic system (without heat loss from the source, the engine and the sink).

Let us suppose that we have heat loss from the hot source through the engine towards the heat sink, Q_L. In this case the heat expense becomes Q_{HS}:

$$Q_{HS} = Q_H + Q_L \tag{5}$$

If Q_H remains the same at the entrance of the converter, we have a new expression of the First Law efficiency given by:

$$\eta_{IL} = \frac{W}{Q_H + Q_L} \tag{6}$$

The consequence is that the non-adiabaticity of the engine diminishes efficiency at a constant work output:

$$\eta_{IL} < \eta_I \tag{7}$$

Similarly, let us suppose that the converter is irreversible so that the entropy balance becomes:

$$\frac{Q_H}{T_{HS}} + \frac{Q_C}{T_{CS}} + \Delta S_i = 0 \tag{8}$$

where ΔS_i is the production of entropy inside the converter during a cycle.

The combination of Equations (8) and (4) with the energy balance allows us to easily find that:

$$\eta_I = \eta_C - \frac{T_{CS}\,\Delta S_i}{Q_H} \tag{9}$$

This proves that the Carnot efficiency η_C is the upper limit of the efficiency for the adiabatic and reversible system (two thermostats and the converter).

2.2. The Curzon–Ahlborn Model [1]

This model also starts from the same point as the one of Equilibrium Thermodynamics. It also considers the Carnot cycle with four thermodynamic processes, but introduces the finite duration of these processes, namely for the one relative to heat transfer at the source and the sink that satisfies:

$$Q = \int_0^\tau \dot{Q}\, dt = \int_0^\tau K'(T_S - T)\, dt \tag{10}$$

where $\dot{Q}$ represents heat transfer rate at time t; K' represents the heat transfer conductance at time t; T_S, T represents respectively source (sink) temperature ,working fluid temperature; τ represents the cycle duration.

It should be noted that special attention should be given to Equation (10), since the integrals are discontinuous due to the fact that the heat transfer at the hot side occurs during the process duration Δt_H, as well as the heat transfer at the cold side, during Δt_C, and are expressed as:

$$Q_H = K'_H(T_{HS} - T_H)\,\Delta t_H > 0 \tag{11}$$

$$Q_C = K'_C(T_{CS} - T_C)\,\Delta t_C < 0 \tag{12}$$

T_H, T_C represents respectively hot side (cold side) working fluid temperature

The mechanical work is expressed through the energy balance as:

$$W = Q_H + Q_C \tag{13}$$

The entropy balance gives a relation between variables T_H, T_C for the **endoreversible case** as follows:

$$\frac{Q_H}{T_H} + \frac{Q_C}{T_C} = 0 \tag{14}$$

Combining Equation (13) with Equations (14) and (4) results in:

$$\eta_I = 1 - \frac{T_C}{T_H} \tag{15}$$

Regarding the optimization of the mechanical work, Curzon and Ahlborn obtained:

$$MAX \ W = \frac{K'_H \, \Delta t_H \cdot K'_C \, \Delta t_C}{K'_H \, \Delta t_H + K'_C \, \Delta t_C} \left(\sqrt{T_{HS}} - \sqrt{T_{CS}} \right)^2 \tag{16}$$

and the corresponding First Law efficiency is given by Equation (2).

If the objective function is the First Law efficiency, we thus recover Carnot's formula.

We would like to add some comments on the present calculation at this point. It is worth noting that $K'_H \cdot \Delta t_H$ and $K'_C \cdot \Delta t_C$ are remarkable quantities, but also that K'_H, K'_C are not directly accessible for experiments, because they are only relative to the transformation part of a cycle (and the transient conditions associated). Indeed, the following expressions, introducing the general physical dimensions, G_i, explain the difference between K' and K:

$$K'_H \, \Delta t_H = K_H \cdot \tau = G_H \tag{17}$$

$$K'_C \, \Delta t_C = K_C \cdot \tau = G_C \tag{18}$$

G's quantities are a kind of invariant for the problem, while this time K_H, K_C correspond to the standard heat transfer conductances, which have been fairly well experimentally correlated in the literature. Additionally, the cycle period is easy to measure.

To conclude, we can see that $Q = G \, (T_S - T)$ imposes a connection between intensity $(T_S - T)$ and extensity Q through G, by analogy with Onsager's appraisal.

Before extending Curzon-Ahlborn's present work, we would like to indicate that they also optimize the power of the engine in the case where adiabatic process durations are short and proportional to isothermal ones, and where the K'_H, K'_C are supposedly equal.

Our proposal regards the general physical dimension G, an extensive coefficient. At first, G_H, G_C were supposed to be constant, but they must remain finite; namely, through K_H, K_C with regard to the system size, and through Δt_H, Δt_C, with regard to times of heat transfer at the hot and cold side, respectively. This is why we propose the unified denomination Finite physical Dimensions Thermodynamics (FDT) as being preferable to Finite Time Thermodynamics (FTT), which only accounts for one aspect (time).

The simplest associated physical dimension constraint is:

$$G_H + G_C = G_T \tag{19}$$

with G_T as a given parameter, and G_H, G_C variables to allocate.

Following the same methodology as before, Equation (16) shows that MAX[MAX (W)] corresponds to the min of $\left(\frac{1}{G_H} + \frac{1}{G_C} \right)$ with the constraint given by Equation (19). This leads easily to the optimal allocation of G_i variables such that:

$$G_H = G_C = \frac{G_T}{2} \tag{20}$$

$$MAX \, [MAX \, (W)] = \frac{G_T}{4} \left(\sqrt{T_{HS}} - \sqrt{T_{CS}} \right)^2 \tag{21}$$

The important and new conclusions are:

- the obtained equipartition of the G_i values in the endoreversible case; and
- the dependence of the maximum mechanical work on only the temperature of the two thermostats, and the allocated G_T invariant.

These results will be used in the following Section 3.

3. Steady-State Modelling

3.1. Heat Transfer Conductance Model

As mentioned in Section 2.2, the modelling with the K_i's and τ is easily accessible for experiments, and corresponds to mean values observed over a great number of cycles. Thus, the characterization of the steady-state functioning of an engine is straightforward and mainly nominal in nature and corresponds to the maximum of power $\dot{W} = W/\tau$. We may also note the corresponding heat rates $\dot{Q}_H = K_H(T_{HS} - T_H)$ at source, $\dot{Q}_C = K_C(T_{CS} - T_C)$ at sink. Consequently, the energy balance is now relative to heat rate and power; and the entropy balance, to entropy rate. The time does not appear explicitly.

There are numerous results associated with steady-state operation of the engine. The author has contributed to these in some papers [12–16].

For endoreversible Carnot steady-state optimization, the same methodology as in Section 2.2 is used (G will be replaced by K) to obtain:

$$K_H = K_C = \frac{K_T}{2} \tag{22}$$

which shows the equipartition of the heat transfer conductances in the endoreversible case, and:

$$MAX\left[MAX\left(\dot{W}\right)\right] = \frac{K_T}{4}\left(\sqrt{T_{HS}} - \sqrt{T_{CS}}\right)^2 \tag{23}$$

All the preceding results are related to the linear form of the heat transfer law (Equation (10)). Studies in the literature [16] report on other heat transfer laws. The most important are:

- The convective heat transfer law:

$$\dot{Q} = K\left(T_S - T\right)^n \tag{24}$$

- The generalized radiative heat transfer law:

$$\dot{Q} = K\left(T_S{}^n - T^n\right) \tag{25}$$

- The phenomenological heat transfer law:

$$\dot{Q} = K\left(\frac{1}{T} - \frac{1}{T_S}\right) \tag{26}$$

with the last law corresponding to Equation (25) with $n = -1$.

All these laws are symmetrical at source and sink. Only a few papers consider different laws at the hot and cold side.

It should be noted that for all these laws, the heat transfer conductance K is not dimensionally standard, and is instead a generalized form of heat transfer conductance. Generally, the corresponding experimental correlations are not given in the literature.

3.2. Refined Heat Exchanger (HEX) Model (Endoreversible Case)

This general model is based on the ε—NTU method (see [16], chapter 2; or chapter 4 of the [16]). Applied to the Carnot engine (Figure 2), it implies:

$$\dot{Q}_H = \varepsilon_H \dot{C}_H (T_{HSi} - T_H) \tag{27}$$

$$\dot{Q}_C = \varepsilon_C \dot{C}_C (T_{CSi} - T_C) \tag{28}$$

where ε_H, (ε_C) is the effectiveness of the Hot, (Cold) HEX; T_{HSi}, (T_{CSi}) is the entrance temperature of the Hot (Cold) external fluid of HEX; $\dot{C}_i = \dot{m}_i \cdot c_{p_i}$ ($i = H$ or C), with $\dot{m}_i$ denoting the mass flow rate at i, and c_{p_i} the specific heat at constant pressure of fluid i.

It is important to note the replacement of the thermostats' limit with a finite heat source and sink, represented respectively by $\dot{C}_H$ and $\dot{C}_C$ heat rates (Finite Speed Thermodynamics).

The methodology remains the same, and by sequential optimization, we obtain for the endoreversible converter case:

$$MAX_1 \ \dot{W} = \frac{\varepsilon_H \dot{C}_H \cdot \varepsilon_C \dot{C}_C}{\varepsilon_H \dot{C}_H + \varepsilon_C \dot{C}_C} \left(\sqrt{T_{HSi}} - \sqrt{T_{CSi}} \right)^2 \tag{29}$$

In the case of finite thermal capacity rate of the source and sink, we can see that G_i introduced in Section 2.1 changes to:

$$G_i = \varepsilon_i \cdot \dot{C}_i \tag{30}$$

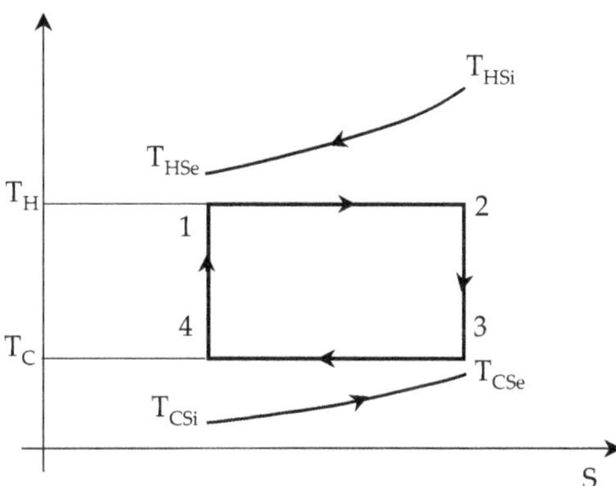

Figure 2. Carnot engine cycle with finite capacity heat rates at source and sink, represented in T-S diagram.

The result expressed by Equation (21) could be applied, but we refined it with the constraint $\varepsilon_H + \varepsilon_C = \varepsilon_T \leq 2$, while supposing $\dot{C}_H$ and $\dot{C}_C$ to be given parameters. We get the optimal allocation of effectiveness as [16] (second sequential optimization relative to effectiveness variables):

$$\varepsilon_C^* = \frac{\sqrt{\dot{C}_H}}{\sqrt{\dot{C}_H} + \sqrt{\dot{C}_C}} \varepsilon_T, \varepsilon_H^* = \frac{\sqrt{\dot{C}_C}}{\sqrt{\dot{C}_H} + \sqrt{\dot{C}_C}} \varepsilon_T, \tag{31}$$

and

$$MAX_2 \; \dot{W} = \varepsilon_T \frac{\dot{C}_H \cdot \dot{C}_C}{\left(\sqrt{\dot{C}_H} + \sqrt{\dot{C}_C}\right)^2} \left(\sqrt{T_{HSi}} - \sqrt{T_{CSi}}\right)^2 \tag{32}$$

As we can see in Equation (32), a new optimization regarding allocation of the heat capacity rates $\dot{C}_i$ is possible with a new added finite dimension constraint, $\dot{C}_C + \dot{C}_H = \dot{C}_T$, and we get:

$$\dot{C}_H = \dot{C}_C = \frac{\dot{C}_T}{2} \tag{33}$$

This corresponds to the equipartition of heat capacity rates in the endoreversible case, while the maximum power output yields:

$$MAX_3 \; \dot{W} = \frac{\varepsilon_T \dot{C}_T}{8} \left(\sqrt{T_{HSi}} - \sqrt{T_{CSi}}\right)^2 \tag{34}$$

A remarkable result is also obtained related to the efficiency at maximum power output for the adiabatic endoreversible system with finite steady-state heat source and sink:

$$\eta_{I\,3} = 1 - \sqrt{\frac{T_{CSi}}{T_{HSi}}} \tag{35}$$

To sum up Section 3, the maximization power for steady-state modelling depends only on the finite physical dimensions (K, ε, $\dot{C}$), and not explicitly on time. This is why we prefer to refer to Finite physical Dimensions Optimal Thermodynamics (FDOT, and not FTT) which is more correct, and corresponds to reality.

Due to lack of space, we have not developed a Finite Speed Thermodynamics (FST) approach (see reference [10]), but it covers similar ground, and corresponds to derivatives with respect to time (as $\dot{C}$): see Section 3.2.

We shall now move on to the explicit influence of time regarding some examples of transient modelling.

4. Transient Modelling

We shall now focus on a Carnot thermo-mechanical engine receiving heat from a source of finite thermal capacity C (case corresponding to a finite energy source).

We shall suppose—as in Sections 2.2 and 3—that the system is without heat loss but we leave the endoreversible configuration of the converter, such that at time t a production of entropy exists which characterized by the converter entropy production rate $\dot{S}_i(t)$ to be specified.

4.1. Model with Perfect Thermal Contact between Finite Heat Capacity Source and Sink, and Irreversible Engine

The scheme of the system is illustrated in Figure 3. The finite source is characterized by $T_C(t)$ associated with sensitive heat to be delivered by the capacity $C = Mc_P$, which is presumed to be constant. This source delivers heat to a Carnot converter producing instantaneous power $\dot{W}(t)$, and emitting heat by direct contact with the environment at T_0. This corresponds, in fact, to a transient version of the Chambadal model [2].

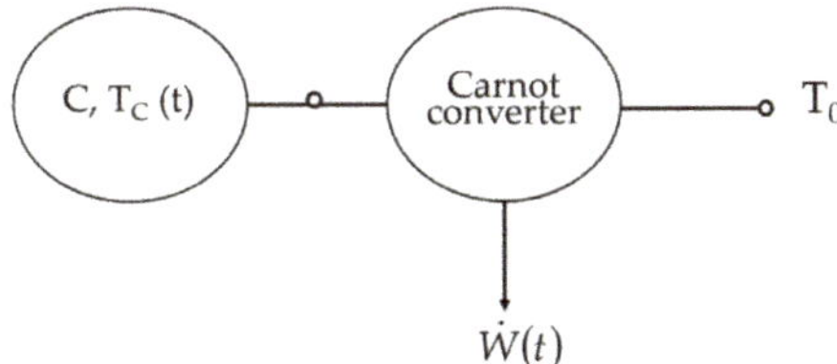

Figure 3. Scheme of a transient version of Chambadal model.

The perfect thermal contact between a finite heat capacity source and the hot side of the Carnot engine implies that the hot side temperature of the engine is identical to $T_C(t)$. By integration over the duration from 0 to final time t_f, it is easy to obtain the work produced in the presence of irreversibility:

$$W = W_{rev} - T_0 \cdot \Delta S_i \tag{36}$$

with: $W_{rev} = C\left(T_{C_0} - T_0\right)\left(1 - \frac{T_0}{\tilde{T}}\right)$, where C denotes the heat capacity of the source,

$$\tilde{T} = \frac{T_{C_0} - T_0}{\ln \frac{T_{C_0}}{T_0}}\text{, entropic mean temperature}$$

$T_{C_0} = T_C(0)$, initial temperature of the source

$$\Delta S_i = \int_0^{t_f} \dot{S}_i(t)\, dt \tag{37}$$

ΔS_i is the entropy production of the irreversible converter during the transient transformation of the system. The entropy production is supposedentirely released to the cycled fluid [17].

Equation (37) clearly indicates that irreversibilities are related to time, but we do not know how, as this depends on the transformation trajectory. However, if t_f goes to infinity (quasi-static transformation), this implies that ΔS_i must converge to zero (reversibility limit). Consequently, the simplest form for ΔS_i is:

$$\Delta S_i = \frac{C_i}{t_f} \tag{38}$$

By combining Equation (36) and (38), it yields:

$$W = W_{rev}\left(1 - \frac{t_{f_0}}{t_f}\right) \tag{39}$$

with $t_{f_0} = \frac{T_0 \cdot C_i}{W_{rev}}$.

The reference time t_{f_0} is the minimum physically acceptable limit for the transient process to deliver work in the presence of irreversibilities (for example due to mechanical friction) [16].

4.2. Optimization of Mean Power over Time

For the transient process, the work appears here as a monotonic increasing function of t_f, with the reversible case as the upper bound.

Regarding $\overline{\dot{W}}$ as the mean power delivered, we get:

$$\overline{\dot{W}} = \frac{W_{rev}}{t_f}\left(1 - \frac{t_{f_0}}{t_f}\right) \tag{40}$$

A derivative with respect to t_f allows us to easily obtain an optimum duration, t_f^*, corresponding to MAX $\overline{\dot{W}}$ in presence of irreversibility, such that:

$$t_f^* = 2\, t_{f_0} \tag{41}$$

$$MAX\ \overline{\dot{W}}\ =\ \frac{W_{rev}}{4\, t_{f_0}}\ =\ \frac{1}{4}\frac{(W_{rev})^2}{T_0\, C_i} \tag{42}$$

The first conclusion is that the maximum mean power of the studied configuration is a decreasing function of the characteristic time t_{f_0}, or of the irreversibility parameter C_i; the two approaches are related (time and entropy).

First Law efficiency corresponds to this maximum power, given by:

$$\eta_I\left(MAX\ \overline{\dot{W}}\right)\ =\ \frac{1}{2}.(1 - \frac{T_0}{\widetilde{T}}) \tag{43}$$

This result completes results obtained and reported in the literature [18,19]. It also shows that the transient conditions are different in terms of efficiency ($\frac{1}{2}.(1 - \frac{T_0}{\widetilde{T}})$) at $MAX\ \overline{\dot{W}}$ than the steady-state one (35). However, the latter is representative of an endoreversible converter (entropy production only as a result of heat transfer). If we take converter irreversibility into account, $\eta_I\left(MAX\ \overline{\dot{W}}\right)$ changes, too, and is related to the irreversibility model, as shown in the transient modelling in [12]. Progress is being made regarding this issue. A first step consists of considering C_i as a function of t_f, instead of as a parameter. Thus, Equation (39) becomes:

$$W\ =\ W_{rev} - \frac{T_0\, C_i\left(t_f\right)}{t_f} \tag{44}$$

Equation (44) allows us to determine t_{f_0} by solving the equation $W = 0$. Thus, the $MAX\ \overline{\dot{W}}$ for t_f^* is the solution of the new equation obtained from the derivative:

$$\frac{2\, T_0\, C_i\left(t_f\right)}{t_f} - \left[W_{rev} + T_0\,\frac{dC_i\left(t_f\right)}{dt_f}\right] = 0 \tag{45}$$

We may note that transient modelling refers explicitly to the time variable through entropy, and that the basic proposed model does not allow sequential optimization related to some specific G dimension. This still needs to be refined.

5. Conclusions and Perspectives

The present paper has shown the insufficiencies of Curzon-Ahlborn's approach, and has proposed solutions to overcome them by:

1. Taking account of heat loss,
2. Taking account of converter irreversibilities,
3. Introducing new intermediate variables, named general physical dimensions: $G_i = K_i'\,\Delta t_i = K_i \cdot \tau$ in the case of the Curzon-Ahlborn model, and $G_i = \varepsilon_i \cdot \dot{C}_i$ in the case of steady-state modelling.

A complete examination of steady-state modelling was proposed in Section 3, and the results were illustrated in the Carnot engine case. A sequential optimization was performed with (1) a first optimization relative to T_i; and (2) a second optimization relative to G_i (with equipartition in the endoreversible configuration of the converter). Additionally, G_i could be $K_i = k_i \cdot A_i$, allowing area A_i to be allocated.

A generalization to a model of HEX was achieved by considering the heat exchanger effectiveness ε_i and, consequently, the heat source and sink of a finite thermal capacity rate. In this case, for the endoreversible converter, the efficiency at maximum power output was expressed by Equation (35), which represents a new upper bound. The irreversible converter case was also studied and found to confirm the Curzon-Ahlborn limit with Equation (44).

The following optimizations regarding Finite Dimensions (ε_i, $\dot{C}_i$) involve equipartition of $\dot{C}_i$ in the endoreversible case, and provide a new expression of the maximum power, given by Equation (34). To summarize the results reported, it emerges that Finite Physical Dimensions Thermodynamics (FDT) allows the optimization of the system sequentially (3 dimensions), and gives new upper bounds not only for endoreversible cases, but also for cases in which an irreversible converter is considered.

Regarding transient modelling, by considering for the first time a finite heat capacity source delivering heat to a Carnot engine (a new version of Chambadal [2] modelling) with perfect thermal contact between capacity and converter, but with internal irreversibility, we obtained the following results:

- MAX (W) was confirmed, corresponding to reversible operation;
- there is a maximum of the mean power output for a finite time transformation with a characteristic time t_{f_0} connected to converter irreversibility. The maximum mean power $MAX\ \overline{W}$ is a decreasing function of t_{f_0}, as well as of irreversibility (C_i). This maximum is obtained for $t_f^* = 2\,t_{f_0}$, and the efficiency that it corresponds to (Equation (43)) is a completely new result (with a value different from the one related to steady-state modelling (see Figure 4, with the theta ratio of the minimum temperature over the maximum temperature).

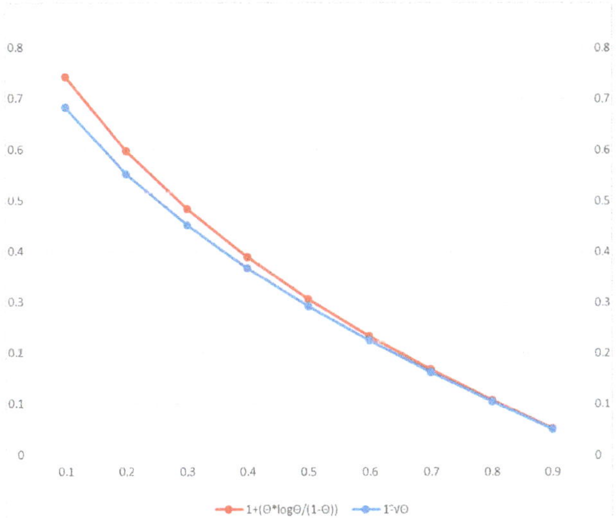

Figure 4. Comparison of efficiency limits at maximum power with the ratio Tmin/TMAX. This result appears more general than the result given in the literature referring to the linear hypothesis of Onsager.

Progress has been made towards a more complete development of these models and results. Particular attention has been given to the combination of various objectives. This approach seems very promising, and gives a new perspective on the subject.

Conflicts of Interest: The author declares no conflict of interest.

References

1. Curzon, F.L.; Ahlborn, B. Efficiency of a Carnot Engine at Maximum Power Output. *Am. J. Phys.* **1975**, *1*, 22–24. [CrossRef]
2. Chambadal, P. *Les Centrales Nucléaires*; A. Colin: Paris, France, 1957. (In French)
3. Novikov, I. The efficiency of atomic power stations (a review). *J. Nucl. Energy* **1958**, *7*, 125–128. [CrossRef]
4. Vaudrey, A.; Lanzetta, F.; Feidt, M.H.B. Reitlinger and the origins of the efficiency at maximum power formula for heat engines. *J. Non-Equilib. Thermodyn.* **2014**, *39*, 199–203. [CrossRef]
5. Chisacof, A.; Petrescu, S.; Costea, M.; Borcila, B. The History of "Nice Radical" and its Importance in Optimizing the Mechanical Work or Power Output of Reversible and Irreversible Cycles. In Proceedings of the Colloque Francophone sur l'Energie, Environnement, Economie et Thermodynamique—COFRET'16, Bucharest, Romania, 29–30 June 2016.
6. Moutier, J. *Éléments de Thermodynamique*; Gautier-Villars: Paris, France, 1872. (In French)
7. Danel, q. Etude Numérique et Expérimentale d'un Cycle de Rankine—Hirn de Faible Puissance Pour larécupération D'énergie. Ph.D. Thesis, Conservatoire National des Arts et Métiers, Paris, France, 12 October 2016. (In French)
8. Rey, J. De l'utilisation de l'énergie des eaux chaudes industrielles et naturelles. *Chaleur et Industrie* **1928**, 217–226. (In French)
9. Serrier, L. *Traité de Physique Industrielle: Production et Utilisation de la Chaleur (Machines à air Chaud)*; G. Masson: Paris, France, 1888. (In French)
10. Petrescu, S.; Costea, M.; Feidt, M.; Ganea, I.; Boriaru, N. *Advanced Thermodynamics of Irreversible Processes with Finite Speed and Finite Dimensions. A Historical and Epistemological Approach, with Extension to Biological and Social Systems*; AGIR Publishing House: Bucharest, Romania, 2015.
11. Calvo Hernandez, A.; Roco, J.M.M.; Medina, A.; Velasco, S.; Guzmán-Vargas, L. The maximum power efficiency $1 - \sqrt{\tau}$: Research, education and bibliometric relevance. *Eur. Phys. J.* **2015**, *224*, 809–823. [CrossRef]
12. Feidt, M.; Costea, M.; Petre, C.; Petrescu, S. Optimization of Direct Carnot Cycle. *Appl. Therm. Eng.* **2007**, *27*, 829–839. [CrossRef]
13. Feidt, M. Optimal Thermodynamics—New Upperbounds. *Entropy* **2009**, *11*, 529–547. [CrossRef]
14. Feidt, M. Thermodynamics of Energy Systems and Processes. *J. Appl. Fluid Mech.* **2012**, *5*, 85–98.
15. Feidt, M. Optimal Use of Energy Systems and Processes. *Int. J. Exergy* **2008**, *5*, 500–531. [CrossRef]
16. Feidt, M. *Thermodynamique Optimale en Dimensions Physiques Finies*; Lavoisier: Paris, France, 2013. (In French); Handbook of Heat Transfer Applications, Chapters 2 and 4 in English.
17. Bizarro, J.P.S. The Thermodynamic Efficiency of Heat Engines with Friction. *Am. J. Phys.* **2012**, *80*, 298. [CrossRef]
18. Feidt, M. Reconsideration of Efficiency of Processes and Systems from a Non-equilibrium Point of View. *Int. J. Energy Environ. Econ.* **2001**, *11*, 31–49.
19. Van den Broeck, C. Thermodynamic Efficiency at Maximum Power. *Phys. Rev. Lett.* **2005**, *95*, 190602. [CrossRef] [PubMed]

Article

Global Efficiency of Heat Engines and Heat Pumps with Non-Linear Boundary Conditions

Per Lundqvist and Henrik Öhman *

Department of Energy Technology, KTH, 100 44 Stockholm, Sweden; per.lundqvist@energy.kth.se
* Correspondence: henrik@hohman.se; Tel.: +46-70-651-8085

Received: 31 March 2017; Accepted: 19 July 2017; Published: 31 July 2017

Abstract: Analysis of global energy efficiency of thermal systems is of practical importance for a number of reasons. Cycles and processes used in thermal systems exist in very different configurations, making comparison difficult if specific models are required to analyze specific thermal systems. Thermal systems with small temperature differences between a hot side and a cold side also suffer from difficulties due to heat transfer pinch point effects. Such pinch points are consequences of thermal systems design and must therefore be integrated in the global evaluation. In optimizing thermal systems, detailed entropy generation analysis is suitable to identify performance losses caused by cycle components. In plant analysis, a similar logic applies with the difference that the thermal system is then only a component, often industrially standardized. This article presents how a thermodynamic "black box" method for defining and comparing thermal efficiency of different size and types of heat engines can be extended to also compare heat pumps of different apparent magnitude and type. Impact of a non-linear boundary condition on reversible thermal efficiency is exemplified and a correlation of average real heat engine efficiencies is discussed in the light of linear and non-linear boundary conditions.

Keywords: global efficiency; energy efficiency; heat engine; heat pump; utilization; Carnot efficiency; comparison; thermal system; cycle analysis

---

1. Introduction

When optimizing a thermal plant, using a heat driven power cycle or a heat pump, practical experience indicates that one seldom have the luxury of choosing the components in any of the systems considered. Instead, the plant designer often needs to choose between preexisting, industrially standardized machines. Such preexisting thermal systems will have characteristics almost according to the designer's preferences, but seldom exactly. Each of the potential thermal system will respond differently to optimizations of the plant. Unless the providers make a complete and unique model available of each potential thermal system, plant optimization has to rely heavily on assumptions. Since such unique models have a tendency to become biased, the problem remains regardless. We therefore propose a different approach to comparing thermal systems on a global level. Global in this approach means that the power cycle or heat pump is treated as a "black box" with global efficiency defined by the real boundary conditions dictated by the plant in which the "black box" operates.

A sound comparison of energy efficiency of thermal systems performing almost similar duty benefits not only from a "black box" approach, but also from a non-dimensional scale and an accurate definition of the reversible energy efficiency of each system. In this article, "thermal systems" means heat engines and heat pumps.

Black box approaches can be defined as independent of technology used. The importance of the black box approach is determined by its purpose. When comparing thermal systems using different cycles, there is no benefit in separating losses internal to the cycle from losses external to the cycle.

By cycle design typical external losses, such as pinch-effects and impact of limited heat exchanger inventories, can be mitigated and are therefore linked to internal cycle losses in various ways. If the purpose instead is to study possible improvement potential of a particular cycle, then conventional second law analysis is highly effective. In such a case, there is no need for a black box approach.

In most practical cases, heat sources as well as heat sinks are finite. Therefore, exit temperatures from the thermal system vary depending on its energy efficiency. A small heat engine, relative to the apparent thermal capacity of source and sink, will operate at a larger temperature difference compared to a large system. Therefore, a comparison between thermal systems of different magnitude requires a measure indicating how small or large each system is relative to the boundary conditions of the heat source and sink. Furthermore, any variation in temperature difference means that an immediate comparison of energy efficiency becomes ambiguous. Instead comparison of energy efficiencies should relate to the reversible energy efficiency of each system, implicitly therefore also to the exit temperatures of the source and sink that would be obtainable using a thermodynamically perfect cycle.

Complexity is added by the fact that many thermal systems operate in environments where apparent heat capacity of heat source and/or heat sink are functions of temperature. We call them "complex", or non-linear. If the apparent heat capacity is constant we call them "linear". In the latter case, analytical formulas can be derived by defining reversible exit temperatures of source and sink as well as reversible energy efficiency. In the complex or non-linear cases, we need a numerical approach.

In the following, we will refer to methods of defining reversible energy efficiency of heat engines and we will propose the same method for comparison of heat pumps. Obviously, refrigeration systems can be compared in a similar manner as heat pumps.

Lorenz [1] defined a type of reversible cycles specifying how the temperatures in finite source and sink changed with transportation of heat through it. With such definition, reversible energy efficiency can be found analytically if the polytropes of the Lorenz cycle constitute equations suitable to integration. In many applications, for example, a heat engine using waste heat recovery of a diesel engine, however the polytropes become difficult to integrate.

Ibrahim & Klein [2] showed a numerical definition of extractable work from a reversible cycle named the "max power cycle". They used multiple, small, Carnot cycles in consecutive order relative to source and sink. In Öhman & Lundqvist [3] the concept of "Integrated Local Carnot Efficiency" $\eta_{C,ll}$ was defined as the reversible thermal efficiency of a max power cycle. Using Figure 1 we can extend the approach to also comprise heat pumps.

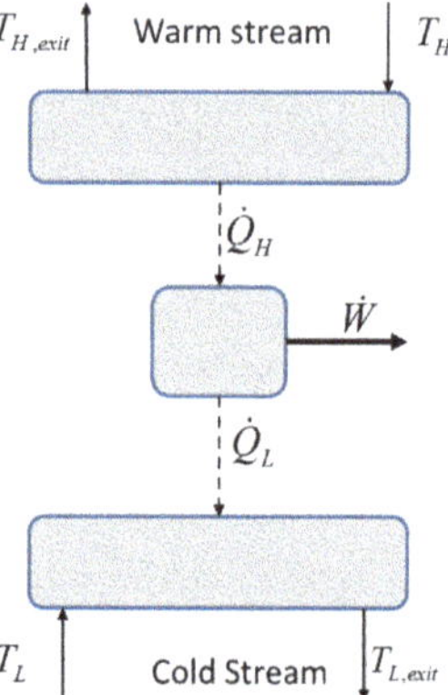

Figure 1. A principal heat engine or, if reversed a heat pump.

Equation (1) shows a definition of energy efficiency, or thermal efficiency, of a reversible heat engine.

$$\eta_{th,s} = \dot{W}_s / \dot{Q}_H \tag{1}$$

Consequently, we may define energy efficiency of a reversible heat pump in Equation (2), operating at the same temperatures.

$$COP_s = 1/\eta_{th,s} \tag{2}$$

As explained in Öhman & Lundqvist [4] we may use the approximation of Curzon–Ahlborn efficiency as of Curzon & Ahlborn [5] in Equation (3), to define the temperature that a source and a sink would equalize to by using a reversible heat engine as in Equation (4), the Curzon–Ahlborn temperature T_{CA}. Note however that, as explained in Öhman [6], using Curzon–Ahlborn efficiency limits the use of Equation (4) to applications with a linear source and sink. A generalization to complex sources and sinks requires the numerical solution of the max power cycle to determine this temperature.

$$\eta_{CA} = 1 - \sqrt{T_L/T_H} \tag{3}$$

We wish to stress that the use of Equation (3) does not imply that the method described is developed according to the tradition of finite-time thermodynamics, FFT, or endo-reversible thermodynamics. It is only used as a simplification used to derive Equation (4) and only applicable to linear boundary conditions. Öhman [6] explains that Equation (3) is incorrect, however the error is small enough to allow its use in Equation (4) for determining Equation (5).

FTT, influenced by Curzon & Ahlborn [5] is a related field of science often focusing on cycle optimization and heat transfer in finite environments. Dong et al. [7] explain a general method to obtain optimal operating points for endo-reversible and irreversible heat engines. Ge et al. [8] explain the advances in finite-time thermodynamics for internal combustion cycle optimization. Feidt [9] explains the development in some traditions of thermodynamic analysis and that FTT is based on the idea of reversible cycle operating with irreversible heat exchange. Feidt [10] explains thermodynamic analysis of reverse cycles, clearly showing that FTT focuses on studying effects on thermal efficiency caused by explicit losses emanating from technical limitations. The black box approach focuses on thermal efficiency as a function of the magnitude of a thermal system relative to the source and sink, without assumptions on specific losses. From that reason, it is natural to use that approach also to study effects of complex boundary conditions. The method explained in this article is explicitly designed for simplified communication of the findings to practitioners.

2. Method

Using Equation (1) and the definition of inverse apparent heat capacity in the source as $\alpha_H = 1/(\dot{m}_H \cdot Cp_H)$ and in the sink as $\alpha_L = 1/(\dot{m}_L \cdot Cp_L)$, Öhman & Lundqvist [4] derived Equation (4) as follows. (Note the printing error in the equation in the reference.)

$$T_{CA} = \frac{T_L + T_H \cdot \alpha_L/\alpha_H \cdot \sqrt{T_L/T_H}}{1 + \alpha_L/\alpha_H \cdot \sqrt{T_L/T_H}} \tag{4}$$

As extensively explained in Öhman [6], we can now define a dimensionless scale, named "utilization" and defined in Equation (5), on which to project the energy efficiency of a heat engine.

$$\psi = \dot{Q}_H / \dot{Q}_H(T_{CA}) = \dot{Q}_H / \dot{Q}_{CA} \tag{5}$$

Note that $\dot{Q}_{CA}$ is only determined by the nature of the source and sink, while $\dot{Q}_H$ is the actual rate of heat entering the heat engine.

Now we can construct a diagram by plotting various characteristic data vs. "utilization" in a dimensionless way, thereby allowing comparison of heat engines of different magnitude relative to the finiteness of heat source and sink. This approach is suggested in Öhman [6] for the comparison of the global energy efficiency of different real heat engines. For the correlation of global energy efficiency of a real heat engine, it is possible to define a "Fraction of Carnot" (*FoC*) as of Equation (6). The *FoC* can be explained as measured, or simulated, using thermal efficiency divided by the ideally possible at the given utilization ψ, where $0 \leq \psi \leq 1$.

$$FoC(\psi) = \eta_{th}/\eta_{C,II}(\psi) = \dot{W}_{real}/\dot{W}_s(\psi) \tag{6}$$

Note that, by referring to boundary conditions of the reversible thermal system, *FoC* is not equivalent to common definitions of exergy efficiency. Note also that $\eta_{C,II}(\psi)$, determined by the numerical max power cycle approach, can be easily validated for linear boundary conditions using equations available in standard literature. Appendix A in Öhman [6] provides explicit expressions for such validation.

3. Results

3.1. Global Efficiency of Heat Pumps

Thermodynamic entities of a reversible heat pump can be viewed as a symmetric mirror of a heat engine operating at the same conditions.

Figure 2 shows temperatures of source and sink for two different thermal systems. On the right side, temperatures of a reversible heat engine are indicated, cooling a hot flow from T_H to $T_{H,exit}$ while heating a cold flow from T_L to $T_{L,exit}$. If $\dot{Q}_H = \dot{Q}_{CA}$ exit temperatures coincide. On the left side, temperatures of a reversible heat pump are indicated, heating a hot flow from $T_{H,exit}$ to T_H while cooling a cold flow from $T_{L,exit}$ to T_L. If $T_{H,exit} \neq T_{L,exit}$ a fictitious heat pump equilibrium temperature can be constructed in the same way as at T_{CA} in Equation (4) for linear boundary conditions and from the numerical model for complex boundary conditions. We can therefore use Equations (5) and (7) to determine utilization.

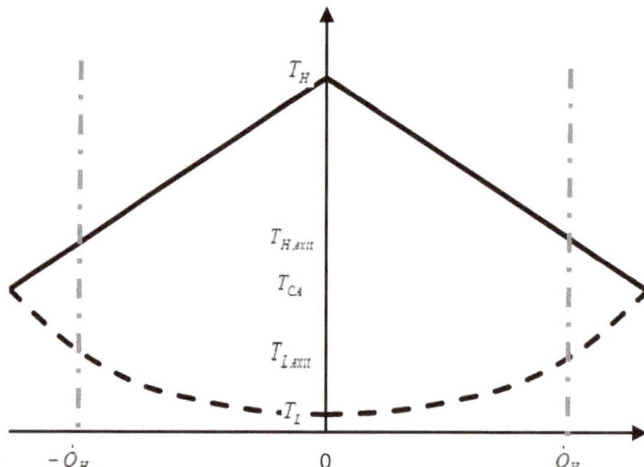

Figure 2. Schematic local temperatures of the hot and the cold streams in a heat pump and a heat engine vs. transported heat (Utilization). Negative rate of heat means heat pump, positive rate of heat means heat engine.

$$\dot{Q}_{CA} = \frac{T_H - T_{H,exit}}{\alpha_H} \tag{7}$$

We can understand that the temperatures in Figure 2 will be symmetric around $\dot{Q}_H = 0$ by the analogy of the max power cycle and its use of multiple subsequent very small Carnot Cycles. Carnot Cycles are reversely applicable to heat engines and heat pumps.

Öhman [6] proposed Equation (8) as the expected correlated global energy efficiency of real low temperature difference heat engines.

$$FoC(\psi) = 0.672 \cdot e^{-0.874 \cdot \psi} \tag{8}$$

Equation (8) is based on measured global energy efficiency for heat engines in a very large range of capacities, temperatures, and technical solutions. Yet, it is remarkably consistent. It allows us very expediently to predict thermal efficiency as well as output power of an undefined real heat engine by using Equation (6), only knowing the characteristics of a particular set of sources and sinks if the utilization is known. The correlation describes an average of historic data and could be seen as normal to industrial capability as of today.

Evaluating heat pump systems can be made in a similar manner as of the above by defining Equation (9) as a Fraction of Carnot for real heat pumps.

$$FoC_{COP}(\psi) = COP_{real}/COP_s(\psi) = \eta_{C,II}(\psi) \cdot COP_{real} \tag{9}$$

A correlation of $FoC_{COP}(\psi)$ has not yet been attempted, but is intended to be performed once enough data has been gathered. It is clear however that global energy efficiency of heat pumps operating in non-identical applications can be systematically compared using the above approach.

3.2. Effects of Complex Boundary Conditions

As explained the simple approximation of Equation (3), and therefore also Equation (4) cannot be applicable if source flow and/or sink flow are complex/non-linear. Utilization can however still be defined by using the local temperatures calculated in the max power cycle. Figure 2 will look significantly different in such a case. Therefore, it is not to be expected that the correlation in Equation (8) is valid for complex boundary conditions.

In reality, complex boundary conditions are common and do require more detailed studies. The approach of the max power cycle, Integrated Local Carnot efficiency and Fraction of Carnot are useful in such studies. The following simplified example can be used to remind us that complex conditions cannot easily be regarded as approximately linear.

3.3. A Demonstrational Example

A demonstrational example can be constructed by applying a heat pump to heat a flow of air by cooling another flow of saturated humid air. A technical application could be a blast drying process or similar and thermally defined as of Table 1.

The specific heat of the cold air is here modeled in two ways; (i) linear, meaning that Cp is constant and that $\alpha_{cooledair}$ is consequently also constant; (ii) complex, meaning that Cp is constant above 34 °C and a function of temperature as of Equation (10) below 34 °C. The reason for the complexity is the condensation of water as the cold air is further cooled below 34 °C. Equation (10) is a polynomial approximation of dh/dT for condensing humid air per mass unit of dry air calculated in the commercial software EES (Engineering Equations Solver) with data on air from Lemmon et al. [11] and data on H_2O from Hyland & Wexler [12].

$$C(T) = 0.0026 \cdot T^3 - 0.0934 \cdot T^2 + 3.5776 \cdot T + 2.6929 \tag{10}$$

Using the set of data from Table 1 we can apply the numerical approach of the max power cycle to obtain the fictitious T_{CA}, output temperature of the cold stream, Utilization, Integrated Local Carnot efficiency and COP for a reversible heat pump for the linear cold stream as well as for the complex cold stream. Net calculated data can be found in Table 2.

Table 1. Input information for a demonstrational example.

Entity	Value	Unit
Hot air flow	3.33	kg/s
Hot air exit temp	80 (dry)	°C
Hot air entry temp	60	°C
Cp (hot air)	1	kJ/kg·K
$\alpha_{heatedairflow}$	0.3	K/kW
Cold air flow	1.67	kg/s
Cold air entry temp	34 (saturated)	°C
Cp (cold air >34 °C)	1	kJ/kg·K
C (cold air <34 °C)	Equation (10)	kJ/kg·K
$\alpha_{cooledair}$ Complex	$C(T_{cooledair})$ Equation (10)	K/kW
$\alpha_{cooledair}$ Linear	0.6	K/kW

Table 2. Calculated data for a reversible blast heater.

Entity	Value	Unit
Hot flow heating	66.7	kW
Reversible COP, complex	8.7	-
Reversible COP, linear	6.5	-
Cold flow exit temp, complex	27	°C
Cold flow exit temp, linear	0.3	°C
Utilization	0.69	-

By comparing the two alternative cold stream characteristics, it becomes obvious that a non-linear, or complex cold stream leads to a difference in reversibly obtainable COP by roughly 30% as compared to assuming a linear cold stream. Figure 3 shows the local temperatures during the two processes. Note that this difference in COP is at reversible conditions. Any measured COP of a real system should be compared to the correct COP, taking the complex nature of the heat sink into account.

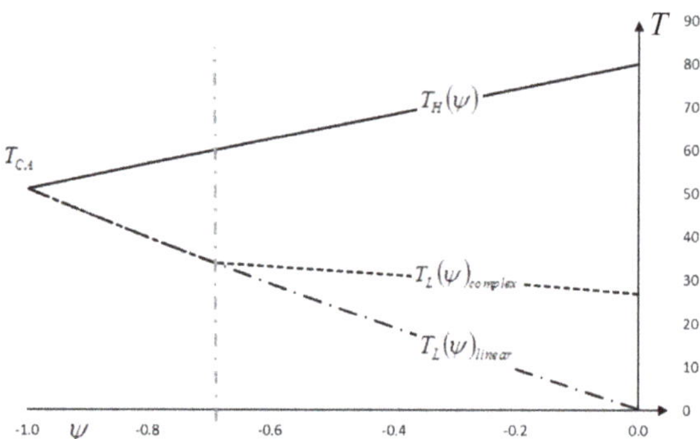

Figure 3. Local temperatures, in °C, of the hot and two alternative cold streams during the two alternative processes as a function of utilization. The calculated processes both start at utilization -0.69 since the heat capacity of the two alternative cold streams are equal when temperature would be above 34 °C. Note that negative utilization indicates a heat pump.

The temperature diagrams in Figure 3 clearly shows why Integrated Local Carnot efficiency must differ between the two processes. Due to the large difference in apparent heat capacity in the cold stream when condensation occurs the necessary temperature lift in the heat pump becomes smaller. This leads to a lower Integrated Local Carnot efficiency and a larger COP for the reversible heat pump.

This example indicates why a correlation similar to Equation (8), but for heat pumps, is not likely to be valid in applications with complex boundary conditions.

4. Discussion

The explained method uses conventional thermodynamic entities to create a dimensionless comparative method for black box energy efficiency. It comprises first as well as second law effects. Other methods can be used but are likely to become more complicated. The numerical approach of the max power cycle provides the benefit of detailed information about local temperatures in source and sink and thereby greatly simplifies the understanding of pinch points and similar effects.

The use of detailed exergy destruction analysis is constructive in identifying and evaluating irreversibility. However, when determining reversible thermal efficiency exergy efficiency becomes meaningless. The approach of Integrated Local Carnot efficiency is a direct implementation of multiple Carnot cycles hence it can be directly derived by assuming zero increase of entropy.

With a similar, systematic black box approach, for heat engines and heat pumps, experiences from one type of thermal system may be useful to understand effects of another.

A particularly interesting question arises when using the concept of utilization for heat engines and heat pumps in a symmetric manner, such as in Figure 2. A research question arising from this discussion is if a correlation, similar to Equation (8) could be found for heat pumps or cooling systems. We suggest researching real heat pump thermal efficiency for further studies using the method described in this article.

5. Conclusions

We have shown that a black box method for investigating efficiency of low temperature heat engines explained in Öhman [6] can be extended to apply also to heat pumps.

The dimensionless scale of utilization can be used for systematic comparison of thermal efficiency of heat engines as well as heat pumps with different magnitudes relative to the source and sink.

The max power cycle approach of Ibrahim & Klein [2] is suitable to determine reversible thermal efficiency of a thermal system with complex boundary conditions.

There may be significant impact of complex boundary conditions on reversible thermal efficiency of a thermal system.

Acknowledgments: This study was partly funded by KTH. KTH covers the costs to publish in open access.

Author Contributions: Henrik Öhman and Per Lundqvist jointly conceived and derived the Method. Henrik Öhman and Per Lundqvist wrote the article.

Conflicts of Interest: The authors declare no conflict of interest. The founding sponsor had no role in the design of the study; in the collection, analyses, or interpretation of data; in the writing of the manuscript, and in the decision to publish the results.

Nomenclature

T_H	Hot flow: entry temp of heat engine/exit temp of heat pump	K
T_C	Cold flow: entry temp of heat engine/exit temp of heat pump	K
$T_{H,exit}$	Hot flow: exit temp of heat engine/entry temp of heat pump	K
$T_{L,exit}$	Cold flow: exit temp of heat engine/entry temp of heat pump	K
$\dot{W}$	Rate of work	W
$\dot{Q}_H$	Rate of heat transferred out from the hot flow	W

$\dot{Q}_L$	Rate of heat transferred into the cold flow	W
$\eta_{th,s}$	Thermal efficiency of a reversible heat engine	-
COP_S	Coefficient Of Performance for a reversible heat pump	-
η_{CA}	Curzon–Ahlborn efficiency	-
T_{CA}	Curzon–Ahlborn temperature	K
$\dot{Q}_{CA}$	Curzon–Ahlborn rate of heat transferred out from the hot flow	W
α_H, α_L	Inverse apparent heat capacity of hot and cold flow	K/W
$\dot{m}_H$, $\dot{m}_L$	Mass flow of hot and cold flow	kg/s
Cp_H, Cp_L	Constant, linear specific heat capacity of hot and cold flow	J/kg·K
$C(T)$	Non-constant, complex specific heat capacity	J/kg·K
h	Specific enthalpy	kJ/kg
ψ	Utilization	-
FoC	Fraction of Carnot for a heat engine	-
FoC_{COP}	Fraction of Carnot for a heat pump	-
$\dot{W}_{real}$	Measured, or simulated rate of work of a thermal system	W
$\dot{W}_s$	Reversible rate of work of a thermal system	W

References

1. Lorenz, H. Beiträge zur Beurteilung von Kuhlmaschinen. *Z. Ver. Dtsch. Ing.* **1894**, *38*, 63–68. (In German)
2. Ibrahim, O.M.; Klein, S.A. Absorption power cycles. *Energy* **1996**, *21*, 21–27. [CrossRef]
3. Öhman, H.; Lundqvist, P. Theory and method for analysis of low temperature driven power cycles. *Appl. Therm. Eng.* **2012**, *37*, 44–50. [CrossRef]
4. Öhman, H.; Lundqvist, P. Comparison and analysis of performance using Low Temperature Power Cycles. *Appl. Therm. Eng.* **2013**, *52*, 160–169. [CrossRef]
5. Curzon, F.L.; Ahlborn, B. Efficiency of a Carnot Engine at Maximum Power Output. *Am. J. Phys.* **1975**, *43*, 22–24. [CrossRef]
6. Öhman, H. Low Temperature Difference Power Systems and Implications of Multiphase Screw Expanders on Organic Rankine Cycles. Ph.D. Thesis, KTH Royal Institute of Technology, Stockholm, Sweden, 2 September 2016.
7. Dong, Y.; El-Bakkali, A.; Descombes, G.; Feidt, M.; Perilhon, C. Association of Finite-Time Thermodynamics and a Bond-Graph Approach for modeling an endoreversible heat engine. *Entropy* **2012**, *14*, 642–653. [CrossRef]
8. Ge, Y.; Chen, L.; Sun, F. Progress in Finite Time Thermodynamic studies for internal combustion engine cycles. *Entropy* **2016**, *18*, 139. [CrossRef]
9. Feidt, M. Optimal thermodynamics—New upperbounds. *Entropy* **2009**, *11*, 529–547. [CrossRef]
10. Feidt, M. Thermodynamics applied to reverse cycle machines, a review. *Int. J. Refrig.* **2010**, *33*, 1327–1342. [CrossRef]
11. Lemmon, E.W.; Jacobsen, R.T.; Pennoncello, S.G.; Freiend, D.G. Thermodynamic properties of air and mixtures of nitrogen, argon, and oxygen from 60 to 2000 K at pressures to 2000 MPa. *J. Phys. Chem. Ref. Data* **2000**, *29*, 331–385. [CrossRef]
12. Hyland, R.; Wexler, A. Formulations for the thermodynamic properties of the saturated phases of H_2O from 173.15 K to 473.15 K. *ASHRAE Trans.* **1983**, *89*, 500–519.

Article

What Is the Real Clausius Statement of the Second Law of Thermodynamics?

Ti-Wei Xue and Zeng-Yuan Guo *

Key Laboratory for Thermal Science and Power Engineering of Ministry of Education, Department of Engineering Mechanics, Tsinghua University, Beijing 100084, China; xtw18@mails.tsinghua.edu.cn
* Correspondence: demgzy@tsinghua.edu.cn

Received: 19 August 2019; Accepted: 19 September 2019; Published: 24 September 2019

Abstract: In this paper, we first analyze the difference between the second law of thermodynamics and the laws in other disciplines. There are some phenomena in other disciplines similar to the Clausius Statement of the second law, but none of them has been accepted as the statement of a certain law. Clausius' mechanical theory of heat, published in the nineteenth century, is then introduced and discussed in detail, from which it is found that Clausius himself regarded "Theorem of the equivalence of the transformation of heat to work, and the transformation of heat at a higher temperature to a lower temperature", rather than "Heat can never pass from a colder to a warmer body without some other change", as the statement of the second law of thermodynamics. The latter is only laid down as the fundamental principle for deriving the theorem of the equivalence of transformations. Finally, based on the theorem of the equivalence of transformations and the average temperature method, a general quantitative relation among the heat, the work, and the temperatures is obtained for arbitrary cycles, which is thus recommended as an alternative mathematic expression of the second law. Hence, the theorem of the equivalence of transformations is the real Clausius Statement of the second law of thermodynamics.

Keywords: second law of thermodynamics; Clausius Statement; theorem of the equivalence of transformations

---

1. Introduction

As with the first law of thermodynamics, the second law of thermodynamics has been verified by countless natural facts. The second law has different statements for different physical phenomena. The classic statements given in most of the literatures mainly include the Clausius Statement and the Kelvin–Planck Statement [1–5]. The Clausius Statement was expressed as "Heat can never pass from a colder to a warmer body without some other change, connected therewith, occurring at the same time", and the Kelvin–Planck Statement as "It is impossible to construct a device that operates in a cycle and produces no other effect than the production of work and the transfer of heat from a single body". The Clausius Statement is more in accord with experience and thus easier to accept, while the Kelvin–Planck Statement provides a more effective means for bringing out second law deductions related to a thermodynamic cycle [3]. Besides, the Carathéodory Statement also occupies an important position in thermodynamics, which is "In the neighbourhood of any arbitrary state of a thermally isolated system, there are states which are inaccessible from that state" [6,7]. It is not more widely used but brings out its essential features in a way that the traditional treatment does not [7]. Since then, various understandings for the statements of the second law came out continuously [8–11], which help to clarify the connotation of the second law from different angles. It is important to emphasize that these statements are logically interlinked, consistent, and equivalent [5,12,13].

However, it is strange that the second law of thermodynamics is quite different from other laws, like Newton's law of motion, Ohm's law of electric conduction, and Fourier's law of heat conduction, etc. There are differences such as; (1) there are many different statements for the same law; (2) it is only a qualitative description of a physical phenomenon, rather than a quantitative relationship between different physical quantities; and (3) some phenomena similar to the Clausius Statement exist in other disciplines. For example, an object can never move from low to high location in a gravity field without some other change, and electrical charges can never move from low to high potential in an electrostatic field without some other change. However, none of them has been accepted as the statement of a certain law. With that in mind, we have reviewed Clausius' mechanical theory of heat, published in the nineteenth century [14,15], and indeed found that he himself only called "Heat can never pass from a colder to a warmer body without some other change, connected therewith, occurring at the same time" a fundamental principle rather than the statement of the second law of thermodynamics.

2. Misunderstanding of the Clausius Statement

2.1. Carnot's Theorem

Before discussing the Clausius Statement, Carnot's work must be first introduced because it is the cornerstone of the second law of thermodynamics [16–19].

Combined with engineering practice experience, Carnot realized that to keep a steam engine running, the working fluid does not only have to absorb heat from the boiler, but also has to release heat to the cooling water in the "condenser" [20]. In other words, when work is generated between two heat reservoirs, there must be a certain amount of heat flowing from the warmer reservoir to the cooler reservoir. The idea of circulation first formed in Carnot's mind [20]. Based on the analogies of "fall of caloric" and "fall of water", Carnot further realized that a certain reversible heat engine cycle always provides the same thermal efficiency for any working fluid. That is to say, the heat transport and the work production obey a certain relationship independent of the nature of the working fluid [15], which is exactly Carnot's theorem.

However, this model does not completely agree with current perceptions [14,15,17,18,21]. Carnot did not yet know the first law of thermodynamics, which states that, when work is produced, the same amount of heat has to be consumed in a reversible cycle. As a result, only a part of the heat is released to the colder body, which is less than the heat from the warmer body. In other words, work cannot be created out of nothing. At that time, thermal efficiency was very low, so the amount of heat absorbed was almost the same as the amount of heat released [20]. In addition, at that time, the backward measurement tools had large measurement errors. Therefore, Carnot could not find the quantity difference between the two parts of heat.

On the positive side, Carnot wondered how the location of the temperature range affects thermal efficiency [22]. He discovered that a given fall of caloric (with a given temperature difference) produces more motive power at inferior than at superior temperatures. In his notation, the thermal efficiency was given by:

$$e = F'(t)dt \tag{1}$$

where $F'(t)$ is a universal function that was called the Carnot function afterwards and t represents the temperature. Unfortunately, because of the cognitive error about heat at that time, Carnot could not determine this function, so he did not know the maximum value of thermal efficiency, not even for an infinitesimal cycle [22]. The Carnot function, however, partly because of its universal character, provided a strong stimulus for further research on the subject. The problem was left open for Clausius to solve twenty five years after Carnot [22].

Even so, Carnot's groundbreaking statement that "The relation between the falling heat and the produced work is independent of the nature of working fluid" is still correct. This laid a foundation for the development of the second law of thermodynamics.

2.2. Clausius Statement

Clausius found early on that Carnot's work about the second law was incomplete and vague [14,15]. He wanted the law of conversion of heat to work in a thermodynamics cycle.

Considering the natural tendency that heat always passes from a warmer to a colder body for eliminating the temperature difference, Clausius presented a well-known proposition that "Heat can never pass from a colder to a warmer body without some other change, connected therewith, occurring at the same time". Unfortunately, it met with much opposition [15]. Muller [22] stated that "this statement, suggestive though it is, has often been criticized as vague". And Muller believed that Clausius himself did not also feel entirely satisfied with this proposition for the reason of it lacking an unequivocal mathematic description [22]. We further find that this proposition had only been laid down as a fundamental principle by Clausius, according to his original literature, "The Mechanical Theory of Heat" [14,15]. In spite of this, regrettably, ever since then, this fundamental principle was usually mistaken for the Clausius Statement of the second law, probably because of its well-understood empirical property.

After Joule's heat-work equivalence theorem was found, Clausius corrected Carnot's theorem successfully [21]. The relation of the heat and the work is still independent of the nature of the working fluid in a reversible cycle, but the heat from the hot source is not the same as the heat going into the cold source, and the difference between them is just the work. In order to get the law of conversion of heat to work, he further considered a reversible thermodynamic cycle as a combination of two kinds of transformations, that is, the transformation of heat to work and the transformation of heat at a higher temperature to a lower temperature [14,15,21]. He thought that "Carnot's theorem actually expressed a relation between the two kinds of transformations, which may be regarded as phenomena of the same nature and are then equivalent" [14]. He called it the theorem of the equivalence of transformations [14,15].

With that in mind, Clausius quantified the two kinds of transformations by giving them "Equivalence-Values". He employed a thermodynamic cycle made up of three isothermal processes and three adiabatic processes [14,15], as shown in Figure 1. Processes f~a, b~c, and d~e are isothermal processes with temperatures t, t_1, and t_2 and heat quantities Q, Q_2, and Q_2, respectively. The other processes are adiabatic processes.

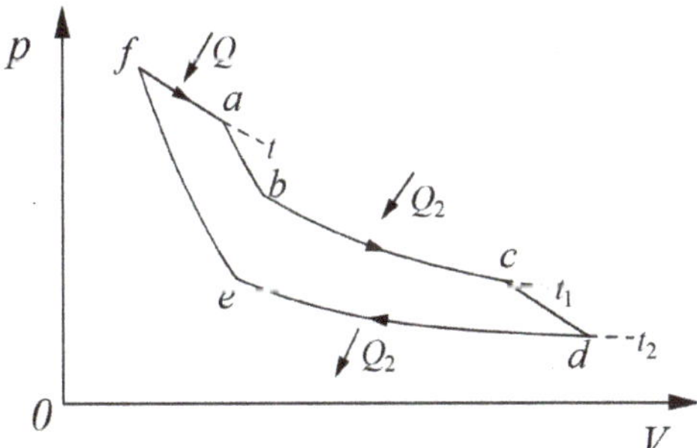

Figure 1. *p-V* diagram of a thermodynamic cycle with three heat reservoirs.

Since the heat quantity Q_2 in process b~c is equal to the heat quantity Q_2 in process d~e, the combination of the two processes could be equivalent to the transformation of heat from a high temperature to a lower temperature, and, thus, process f~a is equivalent to the transformation of heat to work. Clausius considered the transformation of work to heat and, therefore, the transformation of heat from a higher temperature to a lower temperature, as being positive. For the transformation of heat to work, the Equivalence-Value must be proportional to the heat quantity, Q, and must also depend on the temperature, t, which was therefore represented by $-Qf(t)$ [14,15]. In a similar manner, the Equivalence-Value for the heat transformed from a higher temperature to a lower temperature must be proportional to Q_2 and must also depend on the two temperatures t_1 and t_2, which was expressed by $Q_2F(t_1,t_2)$ [14,15]. The two transformations must be equal in magnitude. For the reversible thermodynamic cycle, the sum of the two Equivalence-Values must be zero:

$$-Qf(t) + Q_2F(t_1,t_2) = 0 \tag{2}$$

It is understood that the two transformations in a reversible cycle cancel each other out. Equation (2) is the mathematic expression of the theorem of the equivalence of transformations. Clausius had pointed out that the second law of thermodynamics was called the theorem of the equivalence of transformations in his mechanical theory of heat. Hence, the real Clausius Statement of the second law should be the theorem of the equivalence of transformations.

Harvey [23] stated that Clausius' analysis was brilliant, but the idea of equivalence of transformations is difficult to grasp and is not even mentioned in most thermodynamics textbooks. However, the equivalence of transformations is, we think, of momentous significance for the second law of thermodynamics, as with the equivalence of work and heat for the first law of thermodynamics. The theorem of the equivalence of transformations is a basic theory of classical thermodynamics and it is necessary to be fully aware of it. In the next section, we will see that the Clausius Equality/Inequality and the concept of entropy came from the theorem of the equivalence of transformations [24].

3. Alternative Mathematic Expression of the Second Law

3.1. Clausius Equality/Inequality and Entropy

When t in the cycle in Figure 1 changes to t' and Q thus changes to Q', the following relation appears [14]:

$$-Q'f(t') + Q_2F(t_1,t_2) = 0 \tag{3}$$

If t' is greater than t, then let the thermodynamic cycle, including the temperature, t, work in reverse. The superposition of the two thermodynamic cycles will produce a new Carnot cycle with temperatures t and t'. Then Equation (2) and Equation (3) lead to:

$$\frac{Q}{Q'} = \frac{f(t')}{f(t)} \tag{4}$$

In this new Carnot cycle, Q' can be further divided into two parts; Q'-Q and Q. The first part is regarded as the heat transformed to work and the second part as the heat transformed from a higher temperature to a lower temperature. The following equation can be obtained as with Equation (3):

$$-(Q' - Q)f(t') + QF(t',t) = 0 \tag{5}$$

Based on Equation (4) and Equation (5), the following relation is obtained:

$$F(t',t) = f(t) - f(t') \tag{6}$$

Thus, the Equivalence-Value of the transformation of heat from a higher temperature to a lower temperature can be expressed by:

$$QF(t',t) = Qf(t) - Qf(t') \tag{7}$$

According to Equation (7), the transformation of heat from a higher temperature to a lower temperature is equivalent to the combination of two opposite transformations between heat and work, that is, heat is transformed to work at t', followed by work being transformed to heat at t. Bailyn [21] thus pointed out that, in this sense, the transformation between heat and work is the fundamental transformation. Therefore, for a process with a series of continuous heat reservoirs, the sum N of the Equivalence-Values of all the transformations will be:

$$N = \int f(t)\mathrm{d}Q \tag{8}$$

For a reversible cycle, it can be proved by the fundamental principle that all the transformations must cancel each other out (please refer to Clausius' mechanical theory of heat for details). Thus, the sum of the Equivalence-Values of all the transformations in a reversible cycle must be zero. The integrating factor $f(t)$ is just the reciprocal temperature, $1/T$, that is unique and guaranteed to be a valid integrating factor by the second law [25]. Now we come to the equality:

$$\oint \frac{\mathrm{d}Q}{T} = 0 \tag{9}$$

Clausius called Equation (9) the mathematic expression of the second law of thermodynamics [14,15], which was later named the Clausius Equality.

According to the state axioms, $\mathrm{d}Q/T$ must be the differential of a certain state quantity, which is just entropy, S, that Clausius first discovered, whose differential definition is thus:

$$\mathrm{d}S = \frac{\mathrm{d}Q}{T} \tag{10}$$

For an irreversible thermodynamic cycle, the algebraic sum of all transformations can only be positive [14]. Hence there is an inequality:

$$\oint \frac{\mathrm{d}Q}{T} < 0 \tag{11}$$

where the heat released into a heat reservoir from the working fluid is defined as negative. Afterwards, Equation (11) was called the Clausius Inequality.

The Clausius Equality predicts the entropy flow balance in a reversible cycle, while the Clausius Inequality shows the non-conservation of entropy flow in an irreversible cycle. The Clausius Equality/Inequality, $\oint \delta Q/T \le 0$, that is the combination of Equations (9) and (11) has been widely believed to be the mathematic expression of the second law of thermodynamics [21].

3.2. Alternative Mathematic Expression of the Second Law

As Section 2.1 mentions, Carnot did not find the maximum value of the thermal efficiency because the Carnot function was unknown. Fortunately, Clausius identified the Carnot function and further obtained the thermal efficiency of a Carnot cycle in terms of the theorem of the equivalence of transformations.

For an infinitesimal Carnot cycle, as shown in Figure 2, from the theorem of the equivalence of transformations we can directly derive:

$$-\frac{Q}{T} + \frac{Q - \delta Q}{T - \mathrm{d}T} = 0 \tag{12}$$

Thus, the expression of the thermal efficiency of a Carnot cycle is:

$$e = \frac{\delta Q}{Q} = \frac{1}{T}\mathrm{d}T = \frac{1}{T}\mathrm{d}t \tag{13}$$

Comparing Equation (13) with Equation (1), given by Carnot, the Carnot function can be obtained:

$$F'(t) = \frac{1}{T} \tag{14}$$

So the Carnot function is just the reciprocal of the absolute temperature [14,22]. There is no doubt that Clausius is the first person to determine the Carnot function and the thermal efficiency of a Carnot cycle [22].

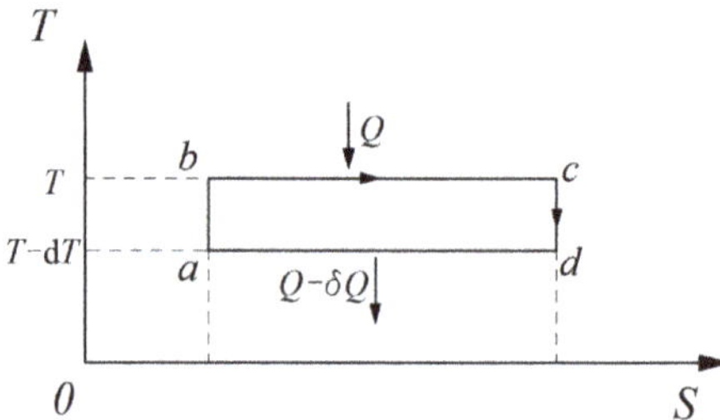

Figure 2. *T-S* diagram of an infinitesimal Carnot cycle.

For a Carnot cycle with two different heat source temperatures, the thermal efficiency can also be derived directly based on the theorem of the equivalence of the transformations. The cycle, as shown in Figure 1, is reconsidered. Because the work quantity, W, is equal to the heat quantity, Q, the Equivalence-Value of transformation of heat to work can be rewritten as $-W/T$. The Equivalence-Value for the heat transformed from the high temperature, T_1, to the low temperature, T_2, is $Q_2 (1/T_2 - 1/T_1)$. The sum of Equivalence-Values of all transformations in the cycle is zero:

$$-\frac{W}{T} + Q_2(\frac{1}{T_2} - \frac{1}{T_1}) = 0 \tag{15}$$

When T changes to T_1, the cycle becomes a Carnot cycle. Thus Equation (15) becomes:

$$-\frac{W}{T_1} + Q_2(\frac{1}{T_2} - \frac{1}{T_1}) = 0 \tag{16}$$

After rearranging Equation (16), we can obtain the expression of thermal efficiency:

$$\eta = \frac{W}{Q_1} = 1 - \frac{T_2}{T_1} \tag{17}$$

which can also be rewritten as a quantity relation between the heat, the work, and the two heat reservoir temperatures:

$$W = Q_1(1 - \frac{T_2}{T_1}) \tag{18}$$

However, for an arbitrary reversible cycle with multiple heat reservoirs, as shown in Figure 3, Equation (18) is no longer valid [26]. One solution to this problem is to employ "the average temperature method" to describe the heat absorption and release processes [3,27,28]. The average temperature method was first proposed by Alefeld [27]. The average temperature, $\overline{T}$, called the entropic average, was defined such that the entropy transfer would be the same if the entire heat transfer occurred at the average temperature [28]:

$$\overline{T} = \frac{Q}{\Delta S} = \frac{Q}{\int \frac{1}{T}dQ} \tag{19}$$

This formulation retains the simplicity of the constant temperature model but makes it applicable to varying temperature processes [28]. Regarding entropy as a variable, the average temperature can be also expressed for a reversible process:

$$\overline{T} = \frac{Q}{\Delta S} = \frac{\int T dS}{\Delta S} \tag{20}$$

Thus, the heat, Q, can be represented by the area under the curve in the T-S diagram. As shown in Figure 3, the entropy value of the working fluid is minimum at point 1 and maximum at point 2. Therefore, process 1-3-2 is the heat absorption process, while process 2-4-1 is the heat release process. The average temperature of the heat absorption process and the average temperature of the heat release process are respectively in an arbitrary reversible cycle:

$$\begin{cases} \overline{T}_1 = \left(\frac{Q}{\Delta S}\right)_{1-3-2} = \dfrac{\int_{1-3-2} TdS}{\Delta S_{12}} = \dfrac{Q_1}{\Delta S_{12}} \\[4mm] \overline{T}_2 = \left(\frac{Q}{\Delta S}\right)_{2-4-1} = \dfrac{\int_{1-4-2} TdS}{\Delta S_{12}} = \dfrac{Q_2}{\Delta S_{12}} \end{cases} \tag{21}$$

where ΔS_{12} is the total entropy change in the heat absorption process or release process, always defined as positive, Q_1 represents the entire exchanged heat of the heat absorption process, equal to the area under the curve 1-3-2 to the S-axis, and Q_2 represents the entire exchanged heat of the heat release process, equal to the area under the curve 2-4-1 to the S-axis. Thus, the quantity relation between the heat and the work for an arbitrary reversible cycle can be represented as:

$$W = Q_1\left(1 - \frac{Q_2}{Q_1}\right) = Q_1\left(1 - \frac{\overline{T}_2 \Delta S_{12}}{\overline{T}_1 \Delta S_{12}}\right) = Q_1\left(1 - \frac{\overline{T}_2}{\overline{T}_1}\right) \tag{22}$$

Equation (22), with the same form as Equation (18), is universally available to any reversible cycle.

As Section 3.1 mentioned, for an irreversible cycle, the algebraic sum of all transformations can only be positive. Therefore, the quantity relation between the heat and the work in an arbitrary irreversible cycle can be obtained:

$$W < Q_1\left(1 - \frac{\overline{T}_2}{\overline{T}_1}\right) \tag{23}$$

Finally, a general form of the quantity relation between the heat and the work for arbitrary cycles comes up:

$$W \le Q_1\left(1 - \frac{\overline{T}_2}{\overline{T}_1}\right) \tag{24}$$

The mathematic formula of a science law usually expresses the quantitative relation among different physical quantities such as Ohm's law and Fourier's law. Equation (24) gives a universal quantitative relation between the heat and the work for any kind of cycle that the Clausius Equality/Inequality does not. Therefore, it could probably be perceived as an alternative mathematic expression of the second law of thermodynamics.

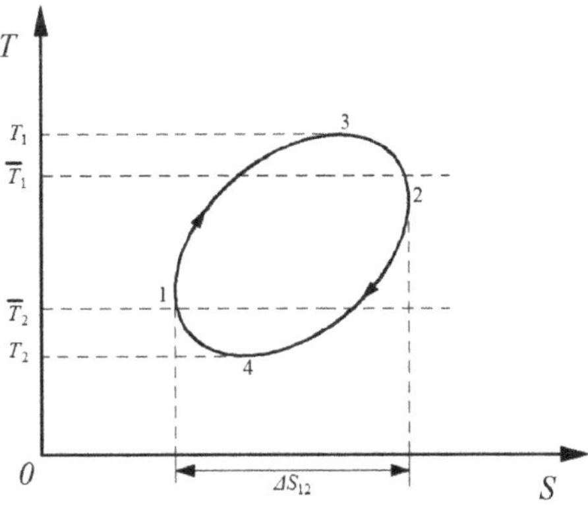

Figure 3. T-S diagram of an arbitrary reversible cycle.

4. Conclusions

(1) Clausius had explicitly proclaimed in his mechanical theory of heat, published in the nineteenth century, that the second law of thermodynamics is "The theorem of the equivalence of transformations", rather than "Heat can never pass from a colder to a warmer body without some other change". The latter, as the description of a natural phenomenon, was only laid down as a fundamental principle. Similar natural phenomena also exist in other disciplines, none of which are yet accepted as the statements of a certain law. Hence, the theorem of the equivalence of transformations is the real Clausius Statement of the second law of thermodynamics.

(2) Precisely because of the theorem of the equivalence of transformations, the quantitative relationship between the heat and work in a Carnot cycle, as well as the Clausius Equality/Inequality, can be derived directly. Finally, the core physical quantity in thermodynamics, entropy, was discovered and defined.

(3) With the method of average temperature we summarize a general quantitative relation among the heat, the work, and the temperatures for any kind of cycles, $W \le Q_1\left(1 - \frac{\overline{T_2}}{\overline{T_1}}\right)$, from the theorem of the equivalence of transformations. It may be regarded as an alternative mathematic expression of the second law of thermodynamics.

Author Contributions: Conceptualization, Z.-Y.G.; funding acquisition, Z.-Y.G.; investigation, T.-W.X. and Z.-Y.G.; methodology, Z.-Y.G.; writing—original draft, T.-W.X.; writing—review and editing, T.-W.X. and Z.-Y.G.

Funding: This research was funded by the Science Fund for Creative Research Groups, grant number 51621062.

Conflicts of Interest: The authors declare no conflict of interest.

References

1. Kondepudi, D.; Prigogine, I. *Modern Thermodymanics: From Heat Engines to Dissipative Structures*; John Wiley & Sons: Hoboken, NJ, USA, 1998; pp. 83, 84, 86.
2. Potter, M.C. *Thermodynamics Demystified*; McGraw-Hill: New York, NY, USA, 2009; pp. 101, 103.
3. Moran, M.J.; Shapiro, H.N. *Fundamentals of Engineering Thermodynamics*, 6th ed.; John Wiley & Sons: Hoboken, NJ, USA, 2010; pp. 239–241, 441.
4. Wark, K. *Advanced Thermodynamics for Engineers McGraw-Hill Series in Mechanical Engineering*; McGraw-Hill: New York, NY, USA, 1995; pp. 41–51.
5. Goodger, E.M. *Principles of Engineering Thermodynamics*, 2nd ed.; Macmillan: London, UK, 1984; pp. 33–34.
6. Carathéodory, C. Untersuchungen über die Grundlagen der Thermodynamik. *Mathematische Annalen* **1909**, *67*, 355–386. (In German) [CrossRef]
7. Adkins, C.J. *Equilibrium Thermodynamics*; Cambridge University Press: Cambridge, UK, 1983; p. 88.
8. Alan, M. A new statement of the second law of thermodynamics. *Am. J. Phys.* **1995**, *63*, 1122.
9. Hillel, A.J. Alternative Statement of the Second Law of Thermodynamics. *Nature* **1973**, *242*, 456–457. [CrossRef]
10. Armstrong, H.L. Statement of the Second Law of Thermodynamics. *Am. J. Phys.* **1960**, *28*, 104–110. [CrossRef]
11. Infelta, P. The Second Law: Statement and Application. *J. Chem. Educ.* **2002**, *79*, 884. [CrossRef]
12. Serrin, J. Conceptual analysis of the classical second laws of thermodynamics. *Arch. Ration. Mech. Anal.* **1979**, *70*, 355–371. [CrossRef]
13. Dunning-Davies, J. Connections between the various forms of the second law of thermodynamics. *Il Nuovo Cimento B* **1969**, *64*, 82–87. [CrossRef]
14. Clausius, R. *The Mechanical Theory of Heat, with Its Applications to the Steam-Engine and to the Physical Properties of Bodies*; John van Voorst: London, UK, 1867; pp. 111–135, 215–256, 327–374.
15. Clausius, R. *The Mechanical Theory of Heat*; Macmillan: London, UK, 1879; pp. 21–38, 69–109, 212–215.
16. Barnett, M.K. Sadi Carnot and the Second Law of Thermodynamics. *Osiris* **1958**, *13*, 31. [CrossRef]
17. Erlichson, H.; Sadi, C. Founder of the Second Law of Thermodynamics. *Eur. J. Phys.* **1999**, *20*, 183–192. [CrossRef]

18. Lemons, D.S.; Penner, M.K. Sadi Carnot's contribution to the second law of thermodynamics. *Am. J. Phys.* **2008**, *76*, 21.

19. Klein, M.J. Carnot's contribution to thermodynamics. *Phys. Today* **1974**, *27*, 23–28. [CrossRef]

20. Fenn, J.B. *Engines, Energy, and Entropy: A Thermodynamics Primer (Global View)*; Freeman: New York, NY, USA, 1982; pp. 93, 104, 131.

21. Bailyn, M. *A Survey of Thermodynamics*; American Institute of Physics: Melville, NY, USA, 1996; pp. 56–65, 74–92, 113–136.

22. Muller, I. *A History of Thermodynamics: The Doctrine of Energy and Entropy*; Springer: Berlin, Germany, 2010; p. 65.

23. Leff, H.S. Thermodynamic entropy: The spreading and sharing of energy. *Am. J. Phys.* **1996**, *64*, 1261. [CrossRef]

24. Cropper, W.H. Rudolf Clausius and the road to entropy. *Am. J. Phys.* **1998**, *54*, 1068–1074. [CrossRef]

25. Weiss, V.C. The uniqueness of Clausius' integrating factor. *Am. J. Phys.* **2006**, *74*, 699. [CrossRef]

26. Tobin, M.C. Engine Efficiencies and the Second Law of Thermodynamics. *Am. J. Phys.* **1969**, *37*, 1115. [CrossRef]

27. Alefeld, G. Efficiency of compressor heat pumps and refrigerators derived from the second law of thermodynamics. *Int. J. Refrig.* **1987**, *10*, 331–341. [CrossRef]

28. Herold, K.E.; Radermacher, R.; Klein, S.A. *Absorption Chillers and Heat Pumps*; CRC Press: Boca Raton, FL, USA, 2016; p. 14.

Article

Performance of a Simple Energetic-Converting Reaction Model Using Linear Irreversible Thermodynamics

J. C. Chimal-Eguia [1,*], R. Paez-Hernandez [2] and Delfino Ladino-Luna [2] and Juan Manuel Velázquez-Arcos [2]

[1]   Centro de Investigación en Computación del Instituto Politécnico Nacional, Av. Miguel Othon de Mendizabal s/n. Col. La Escalera, Ciudad de México, CP 07738, Mexico
[2]   Área de Física de Procesos Irreversibles, Departamento de Ciencias Básicas, Universidad Autónoma Metropolitana, U-Azcapotzalco, Av. San Pablo 180, Col.Reynosa, Ciudad de México, CP 02200, Mexico; rpaez.uam@gmail.com (R.P.-H.); dll@correo.azc.uam.mx (D.L.-L.); jmva@correo.azc.uam.mx (J.M.V.-A.)
*   Correspondence: jchimale@gmail.com

Received: 28 August 2019; Accepted: 17 October 2019; Published: 24 October 2019

Abstract: In this paper, the methodology of the so-called Linear Irreversible Thermodynamics (LIT) is applied to analyze the properties of an energetic-converting biological process using simple model for an enzymatic reaction that couples one exothermic and one endothermic reaction in the same fashion as Diaz-Hernandez et al. (*Physica A*, **2010**, *389*, 3476–3483). We extend the former analysis to consider three different operating regimes; namely, Maximum Power Output (MPO), Maximum Ecological Function (MEF) and Maximum Efficient Power Function (MEPF), respectively. Based on the later, it is possible to generalize the obtained results. Additionally, results show analogies in the optimal performance between the different optimization criteria where all thermodynamic features are determined by three parameters (the chemical potential gap $\Delta = \frac{\mu_1 - \mu_4}{RT}$, the degree of coupling q and the efficiency η). This depends on the election that leads to more or less efficient energy exchange.

Keywords: linear irreversible thermodynamics; maximum power output; maximum ecological Function; maximum efficient power function; enzymatic reaction model

---

1. Introduction

A very interesting problem in non-equilibrium thermodynamics and in the theory of thermodynamics in general, is to determine the efficiency with which energy is exchanged. In fact, in many biological systems, the transfer of energy is of decisive importance. It is well known that all intracellular processes can be studied as chemical reactions of some kind, and that many of the biochemical reactions in living organisms have been seen to be catalyzed by enzymes; there are some good examples where the energetic properties studied are really relevant [1].

Considering the classical ideas of thermodynamics when one wants to analyze biological systems, it is typical to take the free energy of the biological system and convert it into work. For instance, to carry out a transport process or a chemical reaction, it is usual for this type of study to focus on analyzing the energetic properties of such systems. Note, however, that the subject is hard to study from the classical perspective of thermodynamics, since the temperature in many biological systems is homogeneous [2]).

An additional step to analyze the energetic properties of a simple energy converter was given by Curzon and Ahlborn in 1975 [3], who proposed a model which operates between two heat sources with high and low temperatures, T_h and T_c ($T_c < T_h$), respectively. They found an expression for the efficiency at maximum power output given by $\eta_{CA} = 1 - \sqrt{T_c/T_h}$, a result that in principle is independent of the model parameters and only depends on the temperatures of the heat reservoirs; analogously this is what happens with the efficiency for a reversible Carnot cycle $\eta_C = 1 - T_c/T_h$.

Considering the Curzon and Alhbor article, many authors began to introduce different objective functions, among others, such as the ecological function [4], the omega function [5], the efficient power function [6], etc. All of them trying to obtain efficiency and power values mainly for real power plants, but also heat pumps and refrigerators [7,8].

Moreover, it was reported [9–11] that thermal engines show some universality regarding the behavior of the efficiency when it works at the maximum power regime [11], although the analyzed models are different in nature and scale [12–14]. Recently, some thermal engines with kinetic [15–17] and mesoscopic [18] descriptions were published as examples of devices which convert non-thermal energy (mainly chemical energy) into useful work. The importance of these models is that the energy production processes seen in the molecular biological level obey similar principles as those observed in the classical thermal engines [19].

On the other hand, Kedem et al. [20] published in 1965 the first step of a non-equilibrium theory towards a description of linear converters of energy (which would be called Linear Irreversible Thermodynamics, LIT). Since then, many authors have agreed in considering this theory as a basis for the analysis of non-equilibrium systems, (particularly, in biological systems remarkably close to the equilibrium). One of the relevant questions tackled by Kedem et al. at that time was to answer which was the maximal efficiency of the oxidative phosphorylation in an isolated mitochondrion. Kandem et al. obtained some qualitative predictions confirmed by experimental data.

For biological process, for instance, several authors have studied different optimal regimes like Prigogine [21] with his minimum entropy production theorem. Odun and Pinkerton [22] who analyzed the maximum power output regime for various biological systems, Stucki [2] who introduced some optimal criteria to study the optimum oxidative phosphorylation regime, among others [23–28] who have studied many biological energy conversion processes by means of the LIT where some optimum performance regimes have been analyzed.

In this context, we have decided to study the thermodynamical properties of an energetic converting biological process, using for this purpose a simple model for an enzymatic reaction that couples one exothermic and one endothermic reaction in the same fashion described by Diaz-Hernandez et al. [15], but now using the Linear Irreversible Thermodynamics (LIT) for three different operating regimes;namely, Maximum Power Output (MPO), Maximum Ecological Function (MEF) and Maximum Efficient Power Function (MEPF), respectively.

The paper is organized as follows: Section 2 introduces a model and the phenomenological flow equations of a remarkably simple system enzymatic reaction coupled with ATP hydrolysis. Section 3 presents the analysis of the optimal operation regimes in the context of the Linear Irreversible Thermodynamics. Finally, Section 4 gives some concluding remarks.

2. The Model

Consider a simple enzymatic reaction coupled with ATP hydrolysis which might be written as [15]:

$$E + X + ATP \rightleftharpoons [EX] + ADP \rightleftharpoons [EY] + P_i + ADP$$
$$\rightleftharpoons E + Y + P_i + ADP \tag{1}$$

where E represents the enzyme, X is the substrate and Y is the product. Besides $[EX]$ and $[EY]$ are transient complexes of the enzyme with the substrate and the product respectively, ATP corresponds with the Adenosine Triphosphate, ADP is the Adenosine Diphosphate and P_i represents the inorganic phosphate.

Considering the first part of Equation (1), it is possible to obtain the respective reaction velocity, which, according to the mass action law, is given by:

$$\frac{d[E]}{dt} = -k_1[E][X] + k_{-1}[EX] \tag{2}$$

Now, using Arrhenius law, which establishes that,

$$k_1 = F_1 e^{-E_1/RT} \quad \text{and} \quad k_{-1} = F_{-1} e^{-E_{-1}/RT} \tag{3}$$

where, F_1 and F_{-1} are the frequency factors, E_1 and E_{-1} are the activation reaction energies usually expressed in $\frac{cal}{mol}$ and R is the gas constant (expressed in $JK^{-1}mol^{-1}$). Now, from Figure 1, it is clear that both activation energies can be expressed as:

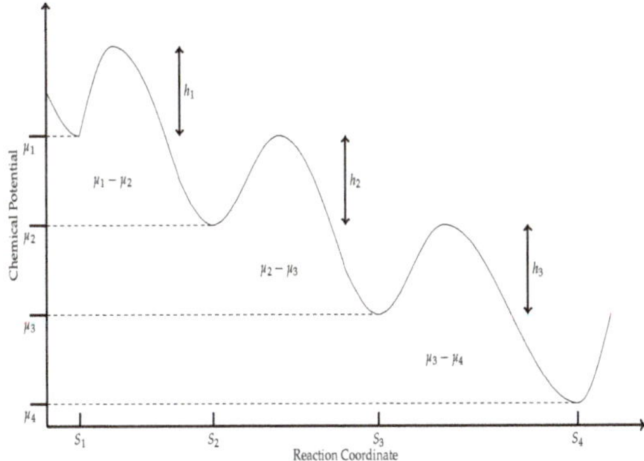

Figure 1. Chemical potential scheme necessary for the reactions in Equation (1) to proceed forward.

$$E_1 = h_1 + \mu_2 - \mu_1 \quad \text{and} \quad E_{-1} = h_1 \tag{4}$$

where μ_i ($i = 1, 2, 3, 4$) is the corresponding chemical potential to the ith state along with the reaction sequence, and h_i ($i = 1, 2, 3$) is the minimum energy required for a collision between molecules to result in a chemical reaction, see Figure 1.

Hence, using Equations (3) and (4) and substituting them into Equation (2), we obtain the net velocities for the three reactions in (1) as:

$$w_i = \xi_i \left(e^{-h_i/RT} - e^{-(h_i + \mu_i - \mu_{i+1})/RT} \right) \tag{5}$$

where $i = 1, 2, 3$ and ξ_i is defined in terms of the molar concentrations and the frequency factors [29]. Then, after some algebra, the above equation can be written as,

$$w_i = A_i \left(1 - e^{-(\mu_i - \mu_{i+1})/RT} \right) \tag{6}$$

where $A_i = \xi_i e^{-h_i/RT}$ (with $i = 1, 2, 3$) and ξ_i is defined as before.

Therefore, by using Equation (6) we can obtain the respective velocities of each reaction as;

$$w_1 = A_1 \left(1 - e^{-(\mu_1 - \mu_2)/RT} \right); \quad w_2 = A_2 \left(1 - e^{-(\mu_2 - \mu_3)/RT} \right); \quad w_3 = A_3 \left(1 - e^{-(\mu_3 - \mu_4)/RT} \right) \tag{7}$$

Now, considering for simplicity that $A_1 = A_3 = A$ and defining $A_2 = \beta A$ (later we will see that β is going to be related to the coupling coefficient q), as Diaz-Hernandez et. al. [15] did in their model using a different approach, we can re-write Equations (7) as;

$$w_1 = A(1 - e^{-(\mu_1 - \mu_2)/RT})$$

$$w_2 = A_2(1 - e^{-(\mu_2 - \mu_3)/RT}) \tag{8}$$

$$w_3 = A(1 - e^{-(\mu_3 - \mu_4)/RT})$$

Since for reactions near equilibrium, the affinity is small, making a Taylor expansion around zero is justified. Keeping this in mind, it is possible to transform Equation (8) into;

$$w_1 = A(\mu_1 - \mu_2)/RT + \mathcal{O}(2)$$

$$w_2 = A_2(\mu_2 - \mu_3)/RT + \mathcal{O}(2) \tag{9}$$

$$w_3 = A(\mu_3 - \mu_4)/RT + \mathcal{O}(2)$$

where we just keep the linear terms in the expansion.

From classical non-equilibrium studies, we can, under suitable conditions, define macroscopic variables locally, as gradients and flux densities. Such variables are called "thermodynamic forces" which drives flux densities often called "fluxes". Following the Onsager formalism [30] we can establish a relation between such forces and fluxes near the steady thermodynamically non-equilibrium regime naming them phenomenological relations [26], given by

$$J_\delta = \sum_\psi L_{\delta\psi} X_\psi \tag{10}$$

where, $L_{\delta\psi}$ are the phenomenological coefficients usually depending on the intensive variables which describes the coupling between two irreversible process δ and ψ, and X_ψ are the respective thermodynamic forces. It is worthwhile to mention that in 1931 Onsager [30] demonstrated that for a system of flows and forces based on an appropriate dissipation function, the matrix of coefficients is symmetrical so that the phenomenological coefficients have the following symmetry relation $L_{\delta\psi} = L_{\psi\delta}$, which affords a considerable reduction in the number of coefficients measured.

Then, taking the above into account, it is possible for our system, to establish two thermodynamic flows J_1 and J_2 for which, we may write the following phenomenological equations;

$$\begin{aligned} J_1 &= L_{11}X_1 + L_{12}X_2 \\ J_2 &= L_{21}X_1 + L_{22}X_2 \end{aligned} \tag{11}$$

where, we are assuming that $L_{12} = L_{21}$.

In the classical equations of chemical kinetics, which are known to describe a chemical process quite precisely, the reaction rates are proportional to the concentrations. On the other hand, phenomenological equations require that the reaction velocity are proportional to the thermodynamic force, which in this case is the Affinity, which is in turn proportional to logarithms of concentration. To remove this inconsistency, we must consider this phenomenological description in the neighborhood of equilibrium when the rate of chemical change is sufficiently slow [31].

According to earlier considerations, if we consider that the driving force for the reaction is the affinity, then close to equilibrium, the chemical flow J_{chem} should be proportional to the force:

$$J_{chem} = L_{ij}\alpha_i = L_{ij}(\mu_i - \mu_j) \tag{12}$$

where L_{ij} are the phenomenological coefficients and $\alpha_i = \mu_i - \mu_j$ is the Affinity. Therefore, assuming that for our chemical reactions the phenomenological relation between fluxes and forces is,

$$J_1 = w_1 + w_2 + w_3 \tag{13}$$

where w_i ($i = 1, 2, 3$) are the net velocities defined in Equation (9). Then, from Equation (13) and substituting Equation (9) in it, we can write;

$$J_1 = A(\mu_1 - \mu_2) + A_2(\mu_2 - \mu_3) + A(\mu_3 - \mu_4) \tag{14}$$

which can be rewritten as:

$$J_1 = A(\mu_1 - \mu_4) + (A_2 - A)(\mu_2 - \mu_3) \tag{15}$$

It is worthwhile to analyze Equation (15). From the scheme in Figure 1, we observe that it corresponds to three sequential equations, all three reactions can be lumped into a single global reaction with free energy change for this reaction as $\Delta G_{TOT} = A(\mu_1 - \mu_4)$. Of the three reactions represented in Figure 1, only in the second one, the consumed energy is used for an interesting purpose; the conversion of the substrate X into the product Y, and the free energy for this reaction is $\Delta G_2 = A(\mu_2 - \mu_3)$. So one possible physical meaning of J_1 is some dissipation-like energy. Then, we can later propose a linear flux–force relation for the enzymatic reaction model as,

$$J_1 = A(\beta - 1)(\mu_2 - \mu_3) + A(\mu_1 - \mu_4)$$
$$\tag{16}$$
$$J_2 = A(\mu_2 - \mu_3) + A(\beta - 1)(\mu_1 - \mu_4)$$

where in the context of linear irreversible thermodynamics we can identify $X_1 = \mu_2 - \mu_3$, $X_2 = \mu_1 - \mu_4$, $L_{12} = L_{21} = A$, and $L_{11} = L_{22} = A(\beta - 1)$, where the parameter β was previously introduced in Equation (7).

We should note that A_i is related to the minimum energy required for a collision between molecules. Thus, this energy could be different for the different stages in the enzymatic reaction causing the phenomenological coefficients L_{ij} to be different, then the degree of coupling q is also different in each stage influencing the thermodynamic properties of the system (for instance, the power output or the entropy production). Considering the latter, we assume that the simplest case is one in which the coefficients A_i are proportional to each other.

Following the concepts of classical thermodynamics, the efficiency function can be defined as [24],

$$\eta = \frac{output}{input} = -\frac{J_1 X_1}{J_2 X_2} \tag{17}$$

From Equation (16), it is possible to substitute J_1 and J_2 into Equation (17), which yields,

$$\eta = \frac{-x(q + Zx)}{qx + 1/Z} \tag{18}$$

where,

$x = \frac{X_1}{X_2} = \frac{\mu_2 - \mu_3}{\mu_1 - \mu_4}$ is the stoichiometric coefficient,

$Z = \sqrt{\frac{L_{11}}{L_{22}}} = 1$ is the phenomenological stoichiometry [2] and

$q = \frac{L_{12}}{\sqrt{L_{11} L_{22}}} = \frac{1}{\beta - 1}$ is the degree of coupling.

Substituting these last expressions in Equation (18) we obtain,

$$\eta = \frac{-x(1 + (\beta - 1)x)}{x + \beta - 1} \tag{19}$$

If we take the special case of complete coupling, i.e., $q = 1$ (from Equation (18) we can notice that for $q = 1, \beta = 2$) in Equation (19), it is easy to observe that,

$$\eta = -x = -\frac{X_1}{X_2} = \frac{\mu_2 - \mu_3}{\mu_1 - \mu_4} \tag{20}$$

which is highly similar to that obtained by Diaz-Hernandez et. al. [15] for a similar model using a different approach. Furthermore, Figure 2 shows the efficiency plotted as a function of the stoichiometric coefficient for various values of q. Note the fast decay of η with the decreasing of q.

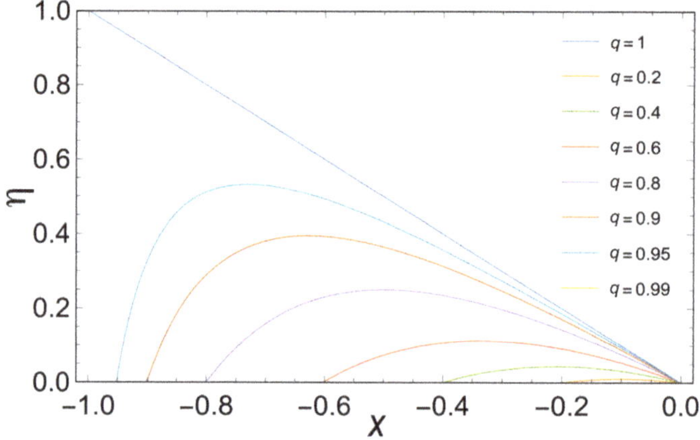

Figure 2. Dependence of the efficiency η on the stoichiometric coefficient x for different values of q from 0.2 up to 1. It is important to remember that the parameter β is related to the degree of coupling as $q = \frac{L_{12}}{\sqrt{L_{11}L_{22}}} = \frac{1}{\beta - 1}$.

3. Optimal Operation Regimes in the Context of the Linear Irreversible Thermodynamics

A very interesting problem in many biological systems is the transfer of energy which is of decisive importance ([23,24,26–28,32]). Caplan et al. [24] studied linear energy converters working in steady states, where they introduced definitions of power output and efficiency, besides the known notion of entropy production rate. Using the definitions of Caplan et al. of power output and efficiency, Stucki [2] analyzed some optimum regimes different from that of minimum entropy production studied before by Prigogine [33]. It has been of special interest in many systems (physical, chemical, biological, etc.) the study of some optimum working regimes for linear energy converters as a manner to understand the diverse ways in which the energy could be transferred [28]. So, let us analyze some of the most representative regimes found in the literature for the present system.

3.1. Maximum Power Output

Using the definitions of Caplan et al. [24] for linear energy converters it is possible to obtain the power output, working in a steady state at constant pressure and temperature, as following;

$$P = -TJ_1 X_1 \tag{21}$$

Taking into account Equation (16) it is possible to substitute them into Equation (21), then we get,

$$P = TL_{22}X_2^2 q^2 v(1 - v) \tag{22}$$

where v is defined as $v = (-L_{11}/L_{12})x$, q and x are defined as in Equation (18) and T is the temperature.

Now, from Equation (19) it is possible to obtain x as a function of η as,

$$x = \frac{-(1+\eta) \pm R}{2(\beta - 1)} \tag{23}$$

with $R = \sqrt{(1+\eta)^2 - 4\eta(\beta - 1)^2}$. Where, we also have considered that $Z = 1$, using the definition given in Equation (18).

If we substitute Equation (23) in Equation (22) we obtain,

$$P = \frac{\Delta^2 TA}{2(\beta - 1)}[(1+\eta) \pm R][\frac{-(1+\eta \pm R)}{2} + 1] \tag{24}$$

where $\Delta = \mu_1 - \mu_4 = X_2$, and A defined as in Equation (6).

It is important to notice that when we take $q = 1$ in Equation (24), we obtain:

$$P = AT\Delta\eta[\Delta(1 - \eta)] \tag{25}$$

which is very similar to the linear approximation of Equation (9) reported by Diaz-Hernandez et al. in Reference [15].

3.2. Maximum Ecological Function

Now, we are going to analyze a regime named ecological. In the context of the Finite Time Thermodynamics [4], the ecological function is defined as,

$$E = P - T\sigma \tag{26}$$

where P is the power output and σ the total entropy production (system plus surroundings) and T the temperature of the cold reservoir. However, in the context of the linear irreversible thermodynamics the ecological function takes the form [27]:

$$E = -TL_{22}X_2^2(2x^2 + 3xq + 1) \tag{27}$$

where again q and x are defined as in Equation (18) and T is the temperature.

Now, taking into account Equation (23) and substitute it into Equation (27) we obtain the ecological function as,

$$E = \frac{\Delta^2 TA}{(1 - \beta)}\{(-\eta \pm R)(\frac{-(1+\eta) \pm R}{2} + 1) \\ +(1 - \beta)^2 - 1\} \tag{28}$$

with $R = \sqrt{(1+\eta)^2 - 4\eta(1 - \beta)^2}$.

It is important to note that when we take $q = 1$ in Equation (28), we obtain:

$$E = AT\Delta(2\eta - 1)[\Delta(1 - \eta)] \tag{29}$$

which is remarkably similar to the linear approximation of Equation (10) reported by Diaz-Hernandez et al. in Reference [15].

As can be seen in Figure 3, the entropy production is a decreasing monotonous function with respect to x, for each value of q. Besides, for a fixed x, we observe that when the entropy production grows q decreases. This could be important because, it seems that could exist a trade-off between the coupling coefficient q and the entropy production σ for a fixed value of x, something pointed out by other authors [15,34].

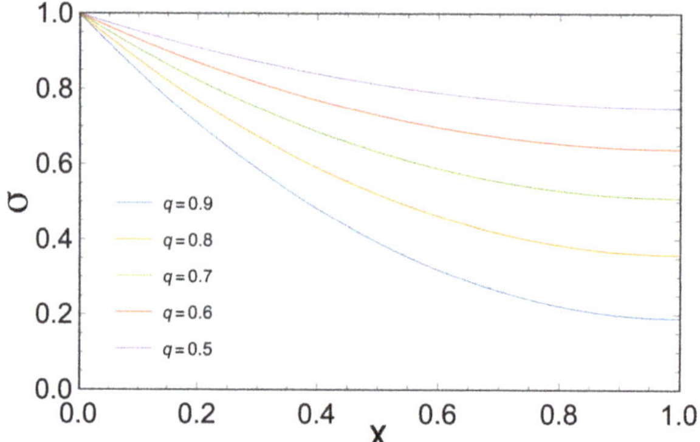

Figure 3. This Figure shows the entropy production σ versus the stoichiometric coefficient x. We observe that σ is a decreasing monotonous function with respect to x, for each value of q.

3.3. Maximum Efficient Power Function

In this section, we present the Maximum Efficient Power regime given by Yilmaz et al. [6] which considers the effects on the design of heat engines, as the multiplication of power by the cycle efficiency, the criteria was successfully applied to the Carnot, Brayton, and Diesel engines, among other systems. From the above, the approach called maximum efficient power in the context of thermal engines is defined as,

$$P_e = \eta P \tag{30}$$

where P is the power output. Maximization of this function provides a compromise between power and efficiency, where the designed parameters at maximum efficient power conditions lead to more efficient engines than those at the maximum power conditions [6].

In the context of the linear irreversible thermodynamics, the power efficient function takes the form,

$$P_e = \frac{\Delta^2 TA}{(1-\beta)} \left\{ \frac{[\frac{-[(1+\eta)\pm R]}{2}]^2[\frac{-(1+\eta)\pm R)}{2}+1]^2}{[\frac{-(1+\eta)\pm R)}{2}+(1-\beta)^2]} \right\} \tag{31}$$

when we take $q = 1$ in Equation (31), we obtain:

$$P_e = AT\Delta\eta^2[\Delta(1-\eta)] \tag{32}$$

Equations (31) and (32) have been obtained considering a new performance criterion, called efficient power, where its maximization leads to a compromise between power and efficiency. In the context of the Linear Irreversible Thermodynamics, the latter is interesting in the sense that we could compare not only the power output, but also the efficiency of the cycle.

3.4. Characteristic Functions vs efficiency

One point of interest in Linear Irreversible Thermodynamics is to obtain information about where the characteristic functions reach their maximum efficiency value; this can be found by means of $\frac{\partial F(\eta,q)}{\partial \eta}$ where $F(\eta, q)$ is any of the three cases considered (i.e., Maximum Power Output (MPO), Maximum Ecological Function (MEF) and Maximum Efficient Power Function (MEPF)).

For the Maximum Power Output (MPO) function, the efficiency which maximizes this function is given by;

$$\eta_{MPO} = \frac{1}{2}\left(\frac{q^2}{2 - q^2}\right) \tag{33}$$

Note some interesting things about Equation (33); first, only when $q = 1$, $\eta_{MPO} = \frac{1}{2}$ in the latest equation. The above is seen clearly in Figure 4 where Power Output (Equation (25)) has been plotted in terms of η and it is observed that the maximum is reached when $\eta = 1/2$. Second, Equation (33) is the same as the one reported by [27], however, the result was obtained here by using a different approach. Third, if we perform a series expansion of Equation (32) in terms of q value around 0, we obtain: $\eta_{MPO} = \frac{q^2}{2}(\frac{1}{2} + \frac{q^2}{4} + \mathcal{O}(q^4))$ this last expression is in some sense equivalent to those founded for heat engines operating between two reservoirs [9–11,19].

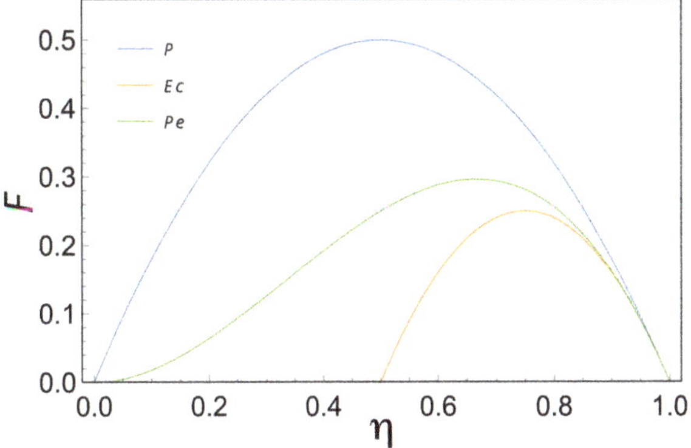

Figure 4. Characteristic Functions (P Maximum Power Output (MPO), Ec Maximum Ecological Function (MEF) and P_e Maximum Efficient Power Function (MEPF)) as a function of the efficiency η. When we take $q = 1$ we can see that Maximum Power Output reaches its maximum at $\eta = 0.5$, the Ecological Function at $\eta = 0.75$, and the Maximum Efficient Power Function at $\eta = 0.6666$.

Now, if we take the Ecological Function, and again we obtain the point where the efficiency maximizes the Ecological Function, we have;

$$\eta_{MEF} = \frac{3}{4}\left(\frac{q^2}{4 - 3q^2}\right) \tag{34}$$

As in the MPO case, when $q = 1$, $\eta_{MEF} = \frac{3}{4}$ in the latest equation, the above is observed clearly in Figure 4 where the Ecological Function (Equation (29)) has been plotted in terms of η and the maximum is reached when $\eta = 3/4$ and $q = 1$, besides Equation (34) is the same as the one reported by [27] but using a different approach. Moreover, performing an expansion of Equation (33) in terms of q we obtain; $\eta_{MEF} = \frac{q^2}{4}(\frac{3}{4} + \frac{9q^2}{16} + \mathcal{O}(q^4))$ and again, this last expansion is similar to those founded in [9–11,19].

Finally, for the case of the Maximum Efficient Power Function (MPEF) when $q = 1$, $\eta_{MPEF} = \frac{2}{3}$. The above is shown clearly in Figure 4 where the Maximum Efficient Power Function (Equation (32)) has been plotted in terms of η and the maximum is reached when $\eta = 2/3$. For that case, if we take the Efficient Power function, and we obtain the point where the efficiency maximizes the Maximum Efficient Power Function as a function of q, we obtain;

$$\eta_{MPEF} = 2*\left(\frac{4}{3q^2} - \frac{2}{3} - \frac{1}{12}\sqrt{-48 + (8 - \frac{16}{q^2})^2}\right) \tag{35}$$

3.5. Loop-Shaped Curves

As Stucki pointed out [2] the coupling coefficient q in real biological systems usually is less than one, this could correspond to some sources of irreversibilities (for example, high thermal conductivity, among others) being different for each case depending on the system. What differs from one engine type to another is the magnitude and source of such irreversibility that gives rise to different power-efficiency curves of this shape [35]. The former has significant differences in the optimal operating conditions for real devices. Hence, looking for loop-shaped power-efficiency curves could aid us in studying the behavior mentioned above.

In order to gain information about the power-efficient curves for the case in which the coupling coefficient q is less than one, we proceed to analyze the loop-shaped curves using the functions of P, E and P_E as a function of η. These are convex functions with respect to x (see Figure 3) when $q=1.$, when q is less than one, we observe that all of them describe a loop-shaped curve with some unusual characteristics such as the maximum obtained in each case for different values of q. The later shows how important the parameter q is since it reflects the behavior of the irreversibilities in the system (see Figure 5–7).

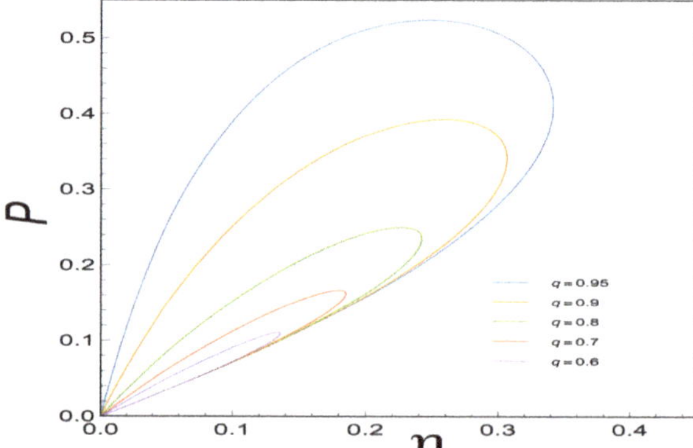

Figure 5. This Figure shows the Power Output P versus the Efficiency η for different values of q. We can observe how the power output produces loop-shaped curves as it is seen in real thermal engines.

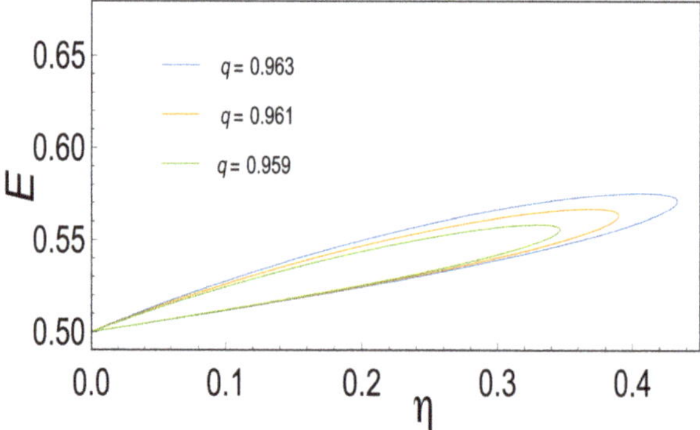

Figure 6. This Figure shows Ecological function P versus the Efficiency η for different values of q. We can observe how the Ecological function produces loop-shaped curves as it is seen in real thermal engines.

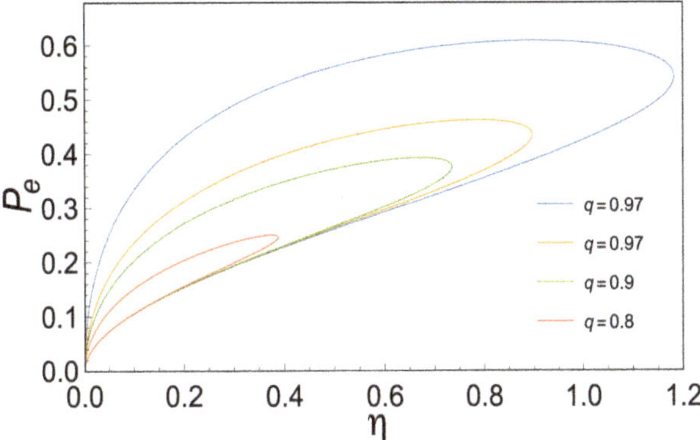

Figure 7. This Figure shows the Efficient Power Function P_e versus the Efficiency η for different values of q. We can observe how the Efficient Power Function produces loop-shaped curves as it is seen in real thermal engines.

4. Concluding Remarks

Many of the intra-cellular processes are studied based on some kind of a chemical reaction. In this work, by using a general model for enzymatic reaction that couples an exothermic with an endothermic reactions (and keeping in mind that most of the biochemical reactions in living organisms are catalyzed by enzymes), we analyzed the efficiency with which energy is exchanged between these reactions, but from the point of view of the non- equilibrium thermodynamics. By using the Linear Irreversible Thermodynamics, it is possible to analyze three different regimes of operation, namely, Maximum Power Output (MPO), Maximum Ecological Function (MEF) and Maximum Efficient Power Function (MEPF).

With that in mind, it is possible to obtain similar expressions for the Power Output and the Ecological Function previously reported by Diaz-Hernandez et al. [15], when the degree of coupling q is equal to one. It is worth mentioning that the studied model is completely based on well known biochemical facts, and in this work, in the context of the Linear Irreversible Thermodynamics, it is possible to generalize the obtained results, where all the thermodynamic features are determined by the chemical potential gap $\Delta = \frac{\mu_1 - \mu_4}{RT}$, the efficiency η, and the degree of coupling q. Moreover, using the same formulation it is possible to add another regime named the Maximum Efficient Power Function in terms of the aforementioned parameters (Equation (31)).

Based on Figure 2, efficiency is a function of the force ratio $x = \frac{X_1}{X_2}$ for various values of q. Note the very rapid fall in η_{max} with the decreasing of q. Again, in the limit $q = 1$, we obtained results comparable to those obtained by Diaz-Hernandez et al. This limit has particularly important thermodynamic implications, since a perfect coupling implies that the flows are not linearly independent. Considering the above, we obtain the efficiency that maximizes the three characteristic functions (MPO, MEF, MPEF), when $q = 1$, we observe that $\eta_{MPO} = 0.5$, $\eta_{MEF} = 0.75$ and $\eta_{MPO} = 0.66$, as is shown in Figure 3.

However, in more realistic scenarios the coupling coefficient q is less than one, for instance Stucki [2] reported an experimental $q_{exp} = 0.95$ for liver mitochondria in male rats. In this case ($q < 1$), we have obtained the efficiency that maximizes the three characteristic functions and also the series expansions in terms of q resembling comparable results yielded by other already cited authors.

Since the three characteristic functions (see Equations (24), (28) and (31)) are determined by three parameters (Δ, q and η) and a variation of $\Delta = \mu_1 - \mu_4$ (this can be achieved assuming a variation of the substrate and the end-product concentrations), the thermodynamic properties could improve because these characteristic functions are proportional to Δ. The latest could be the reason why some biomolecular machines can achieve high speed without sacrificing efficiency [36].

Now, from Equations (33), (34) and (35), it is clear that the efficiency that maximizes some of the characteristic functions is related only to q, so at the end, the thermodynamic properties are related to the degree of coupling providing the basis for comparing different types of coupling in a two-flow system. In other words, the relevant question in many biological situations could be: what is the efficiency with which free energy is exchanged between coupled chemical reactions? (question already made by other authors [26]). Here, the answer is in some sense clear; it depends only on the coupling coefficient and not on the individual phenomenological coefficients, which can be interpreted within the framework of non-equilibrium thermodynamics.

Author Contributions: Conceptualization, J.C.C.-E.; Formal analysis, J.C.C.-E., R.P.-H. and D.L.-L.; Investigation, J.C.C.-E. and R.P.-H.; Methodology, D.L.-L.; Software, J.M.V.-A.; Validation, R.P.-H. and J.M.V.-A.; Writing-original draft, J.C.C.-E.

Funding: Comision de Operacion y Fomento de Actividades Académicas, Instituto Politécnico Nacional: 20196225.

Acknowledgments: The authors want to thanks "Consejo Nacional de Ciencia y Tecnologia" (CONACyT), "Comisión de Operación y Fomento de Actividades Académicas del Instituto Politécnico Nacional" (COFAA-IPN, project number 20196225) and "Estímulos al Desempeño de los Investigadores del Instituto Politécnico Nacional" (EDI-IPN) for the support given for this work.

Conflicts of Interest: The authors declare no conflict of interest.

References

1. Nelson, D.L.; Lehninger, A.L.; Cox, M.M. *Principles of Biochemistry*; Macmillan: Moscow, Russian, 2008.
2. Stucki, J.W. The optimal efficiency and the economic degrees of coupling of oxidative phosphorylation. *Eur. J. Biochem.* **1980**, *109*, 269–283. [CrossRef] [PubMed]
3. Curzon, F.; Ahlborn, B. Efficiency of a Carnot engine at maximum power output. *Am. J. Phys.* **1975**, *43*, 22–24. [CrossRef]
4. Angulo-Brown, F. An ecological optimization criterion for finite-time heat engines. *J. Appl. Phys.* **1991**, *69*, 7465–7469. [CrossRef]
5. Hernández, A.C.; Medina, A.; Roco, J.; White, J.; Velasco, S. Unified optimization criterion for energy converters. *Phys. Rev. E* **2001**, *63*, 037102. [CrossRef] [PubMed]
6. Yilmaz, T. A new performance criterion for heat engines: efficient power. *J. Energy Inst.* **2006**, *79*, 38–41. [CrossRef]
7. Chen, L. *Advances in Finite Time Thermodynamics: Analysis and Optimization*; Nova Publishers: Hauppauge, NY, USA, 2004.
8. De Tomás, C.; Hernández, A.C.; Roco, J. Optimal low symmetric dissipation Carnot engines and refrigerators. *Phys. Rev. E* **2012**, *85*, 010104. [CrossRef]
9. Tu, Z. Efficiency at maximum power of Feynman's ratchet as a heat engine. *J. Phys. A-Math. Theor.* **2008**, *41*, 312003. [CrossRef]
10. Esposito, M.; Lindenberg, K.; Van den Broeck, C. Universality of efficiency at maximum power. *Phys. Rev. Lett.* **2009**, *102*, 130602. [CrossRef]
11. Sanchez-Salas, N.; López-Palacios, L.; Velasco, S.; Hernández, A.C. Optimization criteria, bounds, and efficiencies of heat engines. *Phys. Rev. E* **2010**, *82*, 051101. [CrossRef]
12. Calvo Hernández, A.; Roco, J.; Medina, A.; Sánchez-Salas, N. Heat engines and the Curzon-Ahlborn efficiency. *Rev. Mex. Fis.* **2014**, *60*, 384–389.
13. Schmiedl, T.; Seifert, U. Efficiency at maximum power: An analytically solvable model for stochastic heat engines. *EPL (Europhys. Lett.)* **2007**, *81*, 20003. [CrossRef]
14. Abe, S. Maximum-power quantum-mechanical Carnot engine. *Phys. Rev. E* **2011**, *83*, 041117. [CrossRef] [PubMed]
15. Díaz-Hernández, O.; Páez-Hernández, R.; Santillán, M. Thermodynamic performance vs. dynamic stability in an enzymatic reaction model. *Phys. A* **2010**, *389*, 3476–3483. [CrossRef]
16. Lems, S.; Van der Kooi, H.; de Swaan Arons, J. Thermodynamic optimization of energy transfer in (bio) chemical reaction systems. *Chem. Eng. Sci.* **2003**, *58*, 2001–2009. [CrossRef]

17. Van den Broeck, C.; Kumar, N.; Lindenberg, K. Efficiency of isothermal molecular machines at maximum power. *Phys. Rev. Lett.* **2012**, *108*, 210602. [CrossRef]

18. Golubeva, N.; Imparato, A. Efficiency at maximum power of interacting molecular machines. *Phys. Rev. Lett.* **2012**, *109*, 190602. [CrossRef]

19. Sánchez-Salas, N.; Chimal-Eguía, J.; Ramírez-Moreno, M. Optimum performance for energy transfer in a chemical reaction system. *Phys. A* **2016**, *446*, 224–233. [CrossRef]

20. Kedem, O.; Caplan, S.R. Degree of coupling and its relation to efficiency of energy conversion. *Trans. Faraday Soc.* **1965**, *61*, 1897–1911. [CrossRef]

21. Prigogine, I.; Physicist, C. *Introduction to Thermodynamics of Irreversible Processes*; Interscience Publishers: New York, NY, USA, 1961; Volume 7.

22. Odum, H.T.; Pinkerton, R.C. Time's speed regulator: the optimum efficiency for maximum power output in physical and biological systems. *Am. Scientist* **1955**, *43*, 331–343.

23. Santillán, M.; Angulo-Brown, F. A thermodynamic approach to the compromise between power and efficiency in muscle contraction. *J. Theor. Biol.* **1997**, *189*, 391–398. [CrossRef]

24. Caplan, S.; Essig, A. *Bioenergetics and Linear Nonequilibrium Thermodynamics*; Harvard University Press: Cambridge, MA, USA, 1983.

25. Nelson, P. *Biological Physics*; WH Freeman: New York, NY, USA, 2004.

26. Lebon, G.; Jou, D.; Casas-Vázquez, J. *Understanding Non-Equilibrium Thermodynamics*; Springer: Berlin/Heidelberger, Germany, 2008; Volume 295.

27. Arias-Hernandez, L.A.; Angulo-Brown, F.; Paez-Hernandez, R. First-order irreversible thermodynamic approach to a simple energy converter. *Phys. Rev. E* **2008**, *77*, 011123. [CrossRef] [PubMed]

28. Santillán, M.; Arias-Hernandez, L.A.; Angulo-Brown, F. Some optimization criteria for biological systems in linear irreversible thermodynamics. *Nuovo Cimento-Societa Italiana Di Fisica Sezione D* **1997**, *19*, 99–112.

29. Laidler, K.J. *Reaction Kinetics: Homogeneous Gas Reactions*; Elsevier: Amsterdam, The Netherlands, 2013; Volume 1.

30. Onsager, L. Reciprocal relations in irreversible processes. I. *Phys. Rev.* **1931**, *37*, 405. [CrossRef]

31. Katchalsky, A.; Curran, P.F. *Nonequilibrium Thermodynamics in Biophysics*; Harvard University Press: Cambridge, MA, USA, 1967.

32. Caplan, S. A characteristic of self-regulated linear energy converters the hill force-Velocity relation for muscle. *J. Theor. Biol.* **1966**, *11*, 63–86. [CrossRef]

33. Prigogine, I. *Introduction to Thermodynamics of Irreversible Processes*; Interscience Publishers: New York, NY, USA, 1967; Volume 3

34. Chimal-Eguía, J. Dynamic stability in an endoreversible chemical reactor. *Rev. Mex. Ing. Quim.* **2015**, *14*, 703–710.

35. Gordon, J.; Huleihil, M. General performance characteristics of real heat engines. *J. Appl. Phys.* **1992**, *72*, 829–837. [CrossRef]

36. Wagoner, J.A.; Dill, K.A. Mechanisms for achieving high speed and efficiency in biomolecular machines. *Proc. Nat. Acad. Sci.* **2019**, *116*, 5902–5907. [CrossRef]

Article

Progress in Carnot and Chambadal Modeling of Thermomechanical Engine by Considering Entropy Production and Heat Transfer Entropy

Michel Feidt [1] and Monica Costea [2,*

[1] Laboratory of Energetics, Theoretical and Applied Mechanics (LEMTA), URA CNRS 7563, University of Lorraine, 54518 Vandoeuvre-lès-Nancy, France; michel.feidt@univ-lorraine.fr

[2] Department of Engineering Thermodynamics, University POLITEHNICA of Bucharest, 060042 Bucharest, Romania

* Correspondence: monica.costea@upb.ro or liana5802@yahoo.fr; Tel.: +40-021-402-9339

Received: 11 November 2019; Accepted: 13 December 2019; Published: 16 December 2019

Abstract: Nowadays the importance of thermomechanical engines is recognized worldwide. Since the industrial revolution, physicists and engineers have sought to maximize the efficiency of these machines, but also the mechanical energy or the power output of the engine, as we have recently found. The optimization procedure applied in many works in the literature focuses on considering new objective functions including economic and environmental criteria (i.e., ECOP ecological coefficient of performance). The debate here is oriented more towards fundamental aspects. It is known that the maximum of the power output is not obtained under the same conditions as the maximum of efficiency. This is shown, among other things, by the so-called nice radical that accounts for efficiency at maximum power, most often for the endoreversible configuration. We propose here to enrich the model and the debate by emphasizing the fundamental role of the heat transfer entropy together with the production of entropy, accounting for the external or internal irreversibilities of the converter. This original modeling to our knowledge, leads to new and more general results that are reported here. The main consequences of the approach are emphasized, and new limits of the efficiency at maximum energy or power output are obtained.

Keywords: Carnot engine; optimization; heat transfer entropy; entropy production; new efficiency limits

---

1. Introduction

The origin of the industrial era in the 19th century is strongly connected to thermomechanical engines whose importance in our society is well-known. Their rapid development and evolution have contributed to the foundation of thermodynamics as a new branch of science, mainly concerned with fundamental aspects, as well as numerous applications.

The famous reference case is relative to the Carnot engine [1], but it refers only to an Equilibrium Thermodynamics modelling of a totally reversible engine. Additionally, the work done by Carnot generalizes the concept of efficiency according to the First Law of Thermodynamics starting from a purely mechanical approach [2], and the concept of the Carnot cycle. The interactions between the working fluid and the surroundings during a cycle consist of a mechanical energy exchange, W, and a calorific energy exchange, Q, which are of infinite time duration for reversible conditions. This leads to a quasi-static cycle, whose mean power output of the engine, $\dot{W}$, is zero.

Since (approximately) the 80s, a new approach has been developed including the existence of unavoidable heat transfer irreversibilities (endo-reversible machine [3]), and afterward, internal irreversibilities of the converter were also considered, but less frequently. To take account of these irreversibilities, two main forms are reported in the literature:

- an entropic ratio, I [4];
- a production of entropy, that for a cycle is represented as a parameter, ΔS_I [5].

In the more general case, the entropic ratio could be a function that needs to be specified. This will be considered in this paper.

The main objective encountered in the literature relative to the Non-Equilibrium Thermodynamics approach remains the power output optimization of the engine and the associated conditions (vector of state), as well as the efficiency associated with maximum power.

Numerous studies are concerned with the influence of:

- the form of the heat transfer laws at source and sink [6–10];
- the nature of the sources and sinks (thermostats, fluid flows without phase change) [11,12];
- various objective functions [10,13–16];
- consideration of added constraints [17–25];
- thermal losses or adiabaticity [26–28];
- various irreversibilities (mainly global approaches by considering ΔS_I or I [4]; or introduced by mechanisms, namely solid or fluid energy dissipation) [5].

The present paper proposes a new modelling of an irreversible cycle focused on a necessary condition of the cycle existence, namely the heat transfer entropy. As the most important one for the cycle is the entering heat transfer entropy associated with the heat input in the converter (engine), it will be called hereafter, reference entropy, ΔS. According to the modeling assumptions that consider all cycle processes in the converter as irreversible, the total entropy production over a cycle, ΔS_I is added to this reference heat transfer entropy. The coupling consequence of these two entropies in the analytical analysis of an irreversible cycle leads to new results that are reported here.

This new modelling is developed for the exo-reversible and endo-irreversible Carnot engine configuration, as the first possible extension of the primitive Carnot cycle approach.

Then, the remodeled Chambadal [29] engine configuration is described by adding the influence of the converter internal irreversibility. The optimization approach considers two ways to introduce the converter internal irreversibility, namely the entropic ratio model and the entropic production one, each of them being developed for different dependence functions of the system parameters. The consequences of this addition on the optimized power output are reported, and the limit cases—the endo-reversible Carnot engine and endo-irreversible model—are shown to be recovered.

A discussion of the overall results and an attempt to generalize the present study provide new and important upperbounds for the engine performance.

An ongoing next step will be the application of the same methodology to Curzon-Ahlborn [30] modelling of the Carnot engine by including internal irreversibilities of the converter and considering the importance of the heat transfer entropy.

2. The Endo-Irreversible Carnot Engine

2.1. Cycle Representation in T-S Diagram

The thermomechanical engine is considered as a system composed of a hot bath (heat source), converter (engine) and cold bath (heat sink), as shown in Figure 1. The coupling of the three subsystems is made by the heat transferred from the source to the working fluid in the engine at its hot side, and the heat rejected by the working fluid to the sink at the cold side, respectively.

In the Carnot engine modeling, the source and sink are two infinite bath (thermostats) at temperatures T_{HS} and T_{CS} respectively. The Carnot engine operates upon an irreversible cycle with perfect thermal contact (no temperature difference) with the two bathes. Thus, one can call it an exo-reversible and endo-irreversible engine.

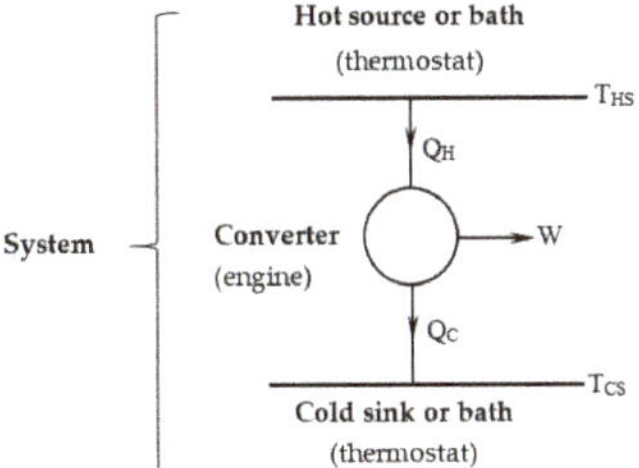

Figure 1. Schematic representation of the thermo-mechanical engine.

The irreversible cycle is identified in Figure 2 from state 1 (beginning of the high temperature isotherm), 2 (exit of the high temperature isotherm and beginning of the adiabatic expansion), 3 (end of the adiabatic expansion and beginning of the low temperature isotherm), and 4 (exit of the low temperature isotherm and beginning of the adiabatic compression, up to state 1). The reversible cycle (marked in red) is also represented in Figure 2 as a starting graphical cycle helping to emphasize the effect of each internal irreversibility considered in the modeling.

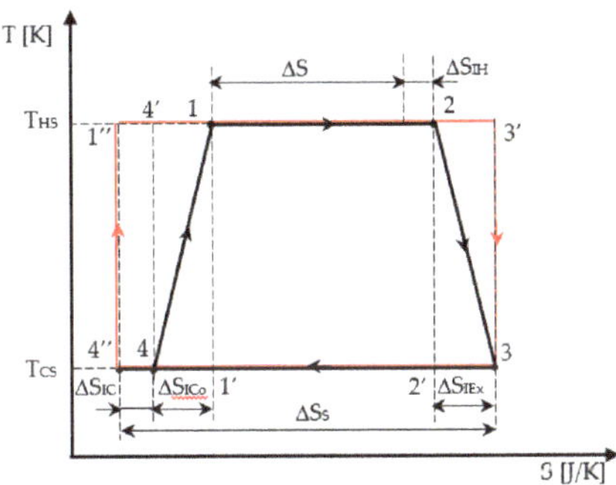

Figure 2. Representation of the Carnot engine cycle with internal irreversibility and the corresponding reversible one (in red) in T-S diagram.

Both adiabatic and isothermal transformations of the cycle are considered irreversible. Thus, the irreversibility on the isothermal processes is represented on the horizontal lines 12 and 34 by the corresponding production of entropy. As the irreversibility on the adiabatic processes in the cycle is also emphasized by entropy production, the graphical representation is associated to a polytropic processes represented in the T-S diagram by a corresponding dependence $T = f(S)$. For simplicity, a linear dependence $T = f(S)$ between states 2 and 3, and 4 and 1, respectively, replaced the real one in the analytical and graphical analysis of the cycle.

Consequently, the production of entropy on a cycle ΔS_I representing the total amount of entropy produced in the universe per cycle (= engine + environment) is expressed as:

$$\Delta S_I = \Delta S_{IH} + \Delta S_{IEx} + \Delta S_{ICo} + \Delta S_{IC} \tag{1}$$

with ΔS_{IH}, production of entropy on the hot isotherm;

ΔS_{IEx}, production of entropy during the adiabatic expansion;

ΔS_{ICo}, production of entropy during the adiabatic compression; and

ΔS_{IC}, production of entropy on the cold isotherm, as shown in Figure 2.

Furthermore, for the hot temperature isotherm, the entropy production ΔS_{IH} corresponds to:

$$\Delta S_H = \Delta S + \Delta S_{IH}, \tag{2}$$

where the reference heat transfer entropy is expressed as:

$$\Delta S = \frac{Q_H}{T_{HS}}, \tag{3}$$

and it equals the entropy change in the hot bath (in module).

Similarly, for the cold temperature isotherm, the entropy production ΔS_{IC} corresponds to:

$$\Delta S_C = \Delta S_S - \Delta S_{IC}, \tag{4}$$

as all production of entropy terms are positive (see Figure 2).

The two entropy productions ΔS_{IH} and ΔS_{IC} correspond to the graphical representation in Figure 3.

One outlines here that the entropy balance over the irreversible cycle is expressed as:

$$\Delta S + \Delta S_I = \Delta S_S = \frac{Q_C}{T_{CS}}, \tag{5}$$

where ΔS_S represents the heat transfer entropy corresponding to the heat transferred to the sink.

Note that the commutations between isothermal and adiabatic processes are assumed to be instantaneous. Moreover, one observes that while the two real isothermal processes are well identified on the diagram, it is not the same for the irreversible adiabatic processes linking state 2 to 3, and then 4 to 1. All the curves inscribed in the rectangle (2-3'-3-2'-2) satisfying the criterion of increasing entropy on the adiabatic expansion process are possible.

Plotting a linear transformation in the *T-S* diagram between state 2 and 3 is not the only physical acceptable solution, because in the frame of Non-Equilibrium Thermodynamics, one only has access to points 2 and 3 representing the adiabatic case.

The same discourse can be renewed for the adiabatic compression between 4 and 1. In short, all the curves inscribed in the rectangle (4-4'-1-1'-4) satisfying the criterion of increasing entropy on the adiabatic process are possible. The linear transformation in the *T-S* diagram between state 4 and 1 is again a possible intermediate solution.

This irreversibility approach is similar to the one of a recent publication [31].

The graphical analysis of the cycle with linearized processes between 23 and 41 illustrated in Figure 2 shows the area representing the corresponding linearized work per cycle W_L as:

$$W_L = \left(\frac{\Delta S_H + \Delta S_C}{2}\right)(T_{HS} - T_{CS}). \tag{6}$$

This 'geometric' work (6) is different from the 'thermodynamic' (or mechanical) work that will be expressed in Section 2.2 by (10) since (6) depends on the linear assumption regarding the representation of the irreversible adiabatic processes between 23 and 41 as done in many references [31]. Nevertheless, it does not reflect the exact physical behavior of the endo-irreversible Carnot heat engine that could be different to the linear one. Thus, only (10), which is based on energy and entropy balances, will provide the correct expression of the mechanical work.

2.2. Analytical Model

Considering the adiabaticity (no thermal loss) and exo-reversibility assumptions for the cycle in this model, the heat transfer at the source and sink are respectively:

$$Q_{HS} = Q_H > 0 \; ; \; Q_{CS} = Q_C < 0. \tag{7}$$

The exo-reversibility of the heat transfer relative to the two thermostats imposes:

$$T_{HS} = T_H \; ; \; T_{CS} = T_C. \tag{8}$$

Thus, the entropy balance on the system can be written as:

$$\frac{Q_{HS}}{T_{HS}} + \Delta S_I = \frac{|Q_{CS}|}{T_{CS}}, \tag{9}$$

where from the expression of the mechanical work per cycle from the First Law when the entropy variation corresponding to the heat transfer converted at the source is considered as reference heat transfer entropy, ΔS, $\Delta S_H = \Delta S$. becomes:

$$W = Q_{HS} - |Q_{CS}| = (T_{HS} - T_{CS})\Delta S - T_{CS}\Delta S_I. \tag{10}$$

The first term of the second equality represents the reversible work, and the second one, the mechanical dissipation (by fluid and solid friction mainly).

The expression of W in (10) marks the presence of the internal entropy production over a cycle ΔS_I that is supposed as a parameter. Its relationship with the ratio method is introduced below, by emphasizing ΔS in (10):

$$W = \Delta S\left[(T_{HS} - T_{CS}) - T_{CS}\frac{\Delta S_I}{\Delta S}\right]. \tag{11}$$

Thus, the ratio $\Delta S_I/\Delta S$ appears in (11) and it corresponds to the irreversibility degree d_I introduced by Novikov [32]. This parameter is also related to an entropic ratio according to Ibrahim's approach [4], which represents the absolute value of the ratio of the entropy change in the cold bath to the entropy change in the hot bath and is given as:

$$I = \frac{\Delta S_S}{\Delta S}. \tag{12}$$

One notes that this ratio can be particularized (i.e., $I_H = \Delta S_S/\Delta S_H$), since it represents a coefficient relative to an entropy variation corresponding to the heat exchanges at the hot and cold sides of the converter. Hence, other assignments could be given (I_C, I_{HS}, I_{CS}). For the adiabatic exo-reversible Carnot engine, this ratio is $I = I_H = I_{HS}$ and moreover, it is easy to show that:

$$I = 1 + d_I = 1 + \frac{\Delta S_I}{\Delta S}, \tag{13}$$

and

$$W = \Delta S[(T_{HS} - T_{CS}) - T_{CS}(I - 1)] = \Delta S[T_{HS} - I{\cdot}T_{CS}], \tag{14}$$

$$\eta_I = \frac{W}{Q_{HS}} = 1 - I\frac{T_{CS}}{T_{HS}}. \tag{15}$$

It results that the internal irreversibility diminishes the mechanical work and cycle efficiency.

It remains to analyze how I or ΔS_I vary with the cycle quantities. It can be reasonably assumed that ΔS_I and (or) I depend on ΔS and (or) $\Delta T_S = T_{HS} - T_{CS}$, as the two temperatures are generally parameters of the problem for the Carnot endo-irreversible engine. Hence, two functions can be proposed, whose general forms (currently unknown in practice) are written as:

$$\Delta S_I = f_{SI}(\Delta T_S, \Delta S) \; ; \; I = f_I(\Delta T_S, \Delta S). \tag{16}$$

If there is an optimum of W, it must satisfy the necessary condition imposed by one of the following equation sets:

$$\frac{\partial W}{\partial \Delta T_S} = \Delta S - T_{CS}\frac{\partial f_{SI}}{\partial \Delta T_S} = 0 \; ; \; \frac{\partial W}{\partial \Delta T_S} = \Delta S\left(1 - \frac{\partial f_I}{\partial \Delta T_S}\right) = 0, \tag{17}$$

$$\frac{\partial W}{\partial \Delta S} = \Delta T_S - T_{CS}\frac{\partial f_{SI}}{\partial \Delta S} = 0 \ ; \quad \frac{\partial W}{\partial \Delta S} = \Delta T_S - T_{CS}\frac{\partial f_I}{\partial \Delta S} = 0. \tag{18}$$

3. The Chambadal Engine

The representation of the Chambadal engine irreversible cycle in Figure 3 shows that in addition to the internal irreversibility of the cycle that was graphically detailed in Section 2.1, the external irreversibility corresponding to the heat transfer at temperature difference at the hot side appears. Since the same components of the internal irreversibility for the endo-irreversible Carnot engine are considered here as well, the cycle representation was simplified by only preserving its shape, and the new irreversibility completing the model was emphasized. Hence, in Figure 3 case (a) corresponds to a heat transfer from a source of constant temperature T_{HS}, while case (b) shows a finite heat capacity source with the inlet temperature T_{HSi}. The temperature difference of the source and sink level, ΔT_S, compared to the cycled fluid one, ΔT_{CF}, that remains the same, is also relevant for the heat input in the two cases.

Note that the external irreversibility at the sink is neglected in this approach by considering the heat rejection to sink occurring at ambient temperature ($T_C = T_{CS} = T_0$).

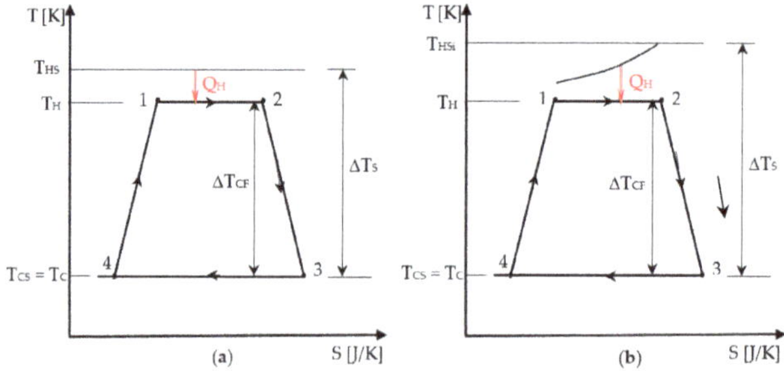

Figure 3. Representation of the associated cycle to the Chambadal engine in T-S diagram: (**a**) source of constant temperature; (**b**) source of finite heat capacity.

The Chambadal engine model is concerned with steady-state operation of the engine and consequently, heat rates and power output are used. We propose here to convert this modeling with reference to a cycle and the related energies-heat and work, thus leading to a different model, the quasi-Chambadal engine.

It involves a thermostat or source of finite heat capacity rate $\dot{C}_H = \dot{m}_H c_{pH}$, with an input temperature T_{HSi}.

Hence, when the source is a thermostat of constant temperature, T_{HS}, the heat transfer at the hot end is expressed as:

$$\dot{Q}_H = U_H A_H (T_{HS} - T_H), \tag{19}$$

with $K_H = U_H \cdot A_H$, heat transfer conductance of the source corresponding to the steady state operation regime.

When the source has a finite heat capacity, it involves a temperature variation of the primary fluid in the corresponding heat exchanger. As the input temperature of the primary fluid T_{HSi} is generally known, by using the effectiveness ε – number of heat transfer units NTU method [27], one gets:

$$\dot{Q}_H = \varepsilon_H \dot{C}_H (T_{HSi} - T_H). \tag{20}$$

Note that the two expressions of $\dot{Q}_H$ averaged on a time duration Δt, allow to recover the caloric energy Q_H, or the mass specific energy (often used) that results by dividing $\dot{Q}_H$ with the mass flow rate of the cycled fluid. Thus, a general expression of the heat transferred at the source is:

$$Q_H = G_H(T_{HSi} - T_H) = T_H \cdot \Delta S, \tag{21}$$

where G_H generally represents a finite physical dimension of the source (thus also of the system) with reference to energy here, its unit being J/K.

3.1. Optimization Approach for the Entropic Ratio Model

The previous equation is completed in this approach by the entropy balance expressed over a cycle as:

$$I_H \frac{Q_H}{T_H} = \frac{Q_C}{T_0}. \tag{22}$$

This leads to the expression of the mechanical work output as:

$$W = G_H(T_{HSi} - T_H)\left(1 - I_H \frac{T_0}{T_H}\right). \tag{23}$$

By taking into account that I_H is a potential function of T_H, the mechanical work maximum value *MaxW* search should satisfy the necessary condition:

$$\frac{\partial W}{\partial T_H} = 0 = -1 + I_H \frac{T_0}{T_H} - (T_{HSi} - T_H)\frac{I_{H,H} T_0 T_H - I_H T_0}{T_H^2}, \tag{24}$$

where $I_{H,H} = \frac{\partial I_H}{\partial T_H}$.

The term rearrangement leads to an equation of second degree or more to be solved:

$$(T_{HSi} - T_H)T_0 T_H I_{H,H} = I_H T_0 T_{HSi} - T_H^2. \tag{25}$$

Possible cases are:

(i) I_H is a constant parameter.

This implies $\partial I_H / \partial T_H - I_{H,H} - 0$.

Thereby, the optimum temperature at the hot-end, *MaxW* and the corresponding cycle efficiency respectively, result as:

$$T_H^* = \sqrt{I_H T_0\, T_{HSi}} \tag{26}$$

$$MaxW = G_H\left(\sqrt{T_{HSi}} - \sqrt{I_H T_0}\right)^2, \tag{27}$$

$$\eta_I(MaxW) = 1 - \sqrt{\frac{I_H T_0}{T_{HSi}}} \tag{28}$$

It is obvious that the rise of I_H parameter results in an increase of the optimal temperature T_H^* and has the opposite effect (decrease) on *MaxW* and the associated cycle efficiency.

(ii) I_H is a linear function of $\Delta T_T = T_H - T_0$ (see Figure 3), with values ≥ 1.

One considers the following expression for I_H:

$$I_H = C_I(T_H - T_0) + 1, \quad with \quad C_I = I_{H,H} \tag{29}$$

By combining (25) with (29), one gets:

$$C_I(T_{HSi} - T_H)T_0 T_H = [C_I(T_H - T_0) + 1]T_0 T_{HSi} - T_H^2. \tag{30}$$

The above equation leads to the expression of the optimum temperature of the cycled fluid at the hot side expressed by:

$$T_H{}^* = \sqrt{T_0\, T_{HSi}}\ .$$ (31)

It results that irreversibility does not affect the temperature $T_H{}^*$, but it does affect the mechanical work and the cycle efficiency, whose expressions become:

$$MaxW = G_H(1 - C_I T_0)\left(\sqrt{T_{HSi}} - \sqrt{T_0}\right)^2,$$ (32)

$$\eta_I(MaxW) = (1 - C_I T_0)\left(1 - \sqrt{\frac{T_0}{T_{HSi}}}\right).$$ (33)

The comparison of (27) and (32) clearly shows the maximum work decrease in case ii) and by a completely different expression. The same happens for the efficiency at maximum work, when comparing (28) and (33).

3.2. Optimization Approach for the Entropic Production Model

The entropy balance takes a similar form to that of Section 2.2, namely:

$$\frac{Q_H}{T_H} + \Delta S_I = \frac{Q_C}{T_C},$$ (34)

with $T_C = T_{CS} = T_0$.

One notes that the production of entropy on the cycle in (34) is not the same with (1) in the endo-irreversible Carnot engine case, since it also considers the production of entropy corresponding to the heat transfer at finite temperature difference (see Figure 3).

The mechanical work is expressed as:

$$W = Q_H\left(1 - \frac{T_{CS}}{T_H}\right) - T_{CS}\Delta S_I.$$ (35)

By using (21) in (35), one gets:

$$W = G_H(T_{HSi} - T_H)\left(1 - \frac{T_{CS}}{T_H}\right) - T_{CS}\Delta S_I.$$ (36)

The term ΔS_I is a potential function of T_H (also increasing one). Therefore, the *MaxW* search should satisfy the necessary condition:

$$\frac{\partial W}{\partial T_H} = 0 = -\left(1 - \frac{T_{CS}}{T_H}\right) + (T_{HSi} - T_H)\frac{T_{CS}}{T_H{}^2} - \frac{T_{CS}\Delta S_{I,H}}{G_H} = 0.$$ (37)

This form is simpler than the previous one, providing an equation of the second degree or more to solve.

The possible cases are as follows:

(i) ΔS_I is a constant parameter.

This implies $\partial S_I/\partial T_H = \Delta S_{I,H} = 0$, which delivers:

$$T_H{}^* = \sqrt{T_{HSi}\, T_{CS}}\ .$$ (38)

The expressions of *MaxW* and $\eta_I(MaxW)$ are straightforward:

$$MaxW = G_H\left(\sqrt{T_{HSi}} - \sqrt{T_{CS}}\right)^2 - T_{CS}\Delta S_I,$$ (39)

$$\eta_I(MaxW) = 1 - \sqrt{\frac{T_{CS}}{T_{HSi}}}\left(1 + \frac{\Delta S_I}{\Delta S}\right). \tag{40}$$

(ii)   A linear dependence of ΔS_I on the temperature difference $\Delta T_T = T_H - T_{CS}$ is considered.

We assume that:

$\Delta S_I = 0$ if $\Delta T_T = 0$;

$\Delta S_I = \Delta S_{IS}$ (new parameter) if $\Delta T_T = \Delta T_S = T_{HSi} - T_{CS}$ (see Figure 3b),

where from:

$$\Delta S_I = \Delta S_{IS}\frac{\Delta T_T}{\Delta T_S} = \Delta S_{IS}\frac{T_H - T_{CS}}{T_{HSi} - T_{CS}} = C_{IS}(T_H - T_{CS}). \tag{41}$$

The mechanical work optimization imposes:

$$\frac{\partial W}{\partial T_H} = 0 = G_H\left(\frac{T_{HSi}\,T_{CS}}{T_H{}^2} - 1\right) - \frac{\Delta S_{IS}}{\Delta T_S}T_{CS}, \tag{42}$$

which leads to:

$$T_H{}^* = \sqrt{\frac{T_{HSi}\,T_{CS}}{1 + \dfrac{T_{CS}}{\Delta T_S}\dfrac{\Delta S_{IS}}{G_H}}} . \tag{43}$$

It is well known that internal irreversibility always decreases W. Hence, (41) is valuable by emphasizing in the denominator two factors that modify the optimal temperature $T_H{}^*$:

- the first, an intensive one: $\frac{T_{CS}}{T_{HSi} - T_{CS}}$;
- the second, an extensive one: $\frac{\Delta S_{IS}}{G_H}$.

The last factor is not only related to the converter through ΔS_{IS}, but also to the hot source by the finite physical dimension G_H (coupling).

This shows the interest of an entropic description of the system. Thus, the corresponding maximum power output is given by:

$$MaxW = G_H\left(\sqrt{T_{HSi}} - \sqrt{I{\cdot}T_0}\right)^2, \tag{44}$$

with

$$I = \frac{T_0}{\Delta T_S}{\cdot}\frac{\Delta S_{IS}}{G_H}. \tag{45}$$

The associated first law efficiency is:

$$\eta_I(MaxW) = \frac{\left(1 - \sqrt{\dfrac{I{\cdot}T_0}{T_{HSi}}}\right)^2}{1 - \sqrt{\dfrac{T_0}{I{\cdot}T_{HSi}}}}, \tag{46}$$

3.3. Quasi-Chambadal Model with Added Heat Transfer Entropy Constraint and Production of Entropy

This optimization is based on (21) from which the temperature T_H is expressed as:

$$T_H = \frac{G_H}{G_H + \Delta S}T_{HS}, \tag{47}$$

with ΔS, reference heat transfer entropy associated to the heat input. This result is an important one, since it illustrates the coupling between the source (G_H) and the converter (ΔS).

The expression of the mechanical work W can be easily derived as a function of the newly chosen variable ΔS:

$$W = \left[\frac{T_{HSi}\,G_H}{G_H + \Delta S} - T_0\right]\Delta S - T_0\,\Delta S_I. \tag{48}$$

This optimum with respect to ΔS corresponds to a converter adapted to the hot source. It results that W becomes maximum when ΔS_I is related to ΔS (to be specified) by the following function:

$$\frac{\partial W}{\partial \Delta S} = 0 = \left(\frac{G_H}{G_H + \Delta S}\right)^2 T_{HSi} - T_0 - T_0 \frac{\partial \Delta S_I}{\partial \Delta S} \, . \tag{49}$$

Following different assumptions, one can consider:

(i) the presence of the Novikov's irreversibility degree [9], that is expressed by:

$$\frac{\partial \Delta S_I}{\partial \Delta S} = d_I. \tag{50}$$

In this case, one gets at the optimum:

$$\Delta S^* = G_H \left[\frac{\sqrt{T_{HSi}}}{\sqrt{T_0 \, (1 + d_I)}} - 1 \right]. \tag{51}$$

It results that the optimum heat transfer entropy ΔS^* diminishes with increasing irreversibility d_I, and the same goes for $MaxW$ and the cycle efficiency:

$$MaxW = G_H \left(\sqrt{T_{HSi}} - \sqrt{T_0 \, (1 + d_I)} \right)^2, \tag{52}$$

$$\eta_I(MaxW) = 1 - \sqrt{\frac{T_0 \, (1 + d_I)}{T_{HSi}}}, \tag{53}$$

where $d_I \le (T_{HSi} - T_0)/T_0$.

(ii) when ΔS_I is a constant parameter, one finds similarly:

$$\Delta S^* = G_H \left[\frac{\sqrt{T_{HSi}}}{\sqrt{T_0}} - 1 \right], \tag{54}$$

and

$$MaxW = G_H \left(\sqrt{T_{HSi}} - \sqrt{T_0} \right)^2 - T_0 \, \Delta S_I, \tag{55}$$

$$\eta_I(MaxW) = 1 - \sqrt{\frac{T_0}{T_{HSi}}} \left(1 + \frac{\sqrt{T_0}}{\sqrt{T_{HSi}} - \sqrt{T_0}} \frac{\Delta S_I}{G_H} \right). \tag{56}$$

(iii) ΔS_I may also depend of T_H in a linear form, as in Section 3.2, case ii), namely:

$$\Delta S_I = d_{IS} \frac{T_H - T_0}{T_{HSi} - T_0} \Delta S, \tag{57}$$

from which one gets:

$$W = (T_H - T_0) \Delta S \left(1 - T_0 \frac{d_{IS}}{T_{HSi} - T_0} \right), \tag{58}$$

where $d_{IS} \le (T_{HSi} - T_0)/T_0$ is the parameter of entropic proportionality corresponding to the irreversibility degree of Novikov [24], but relative to the system composed of a source and converter (the heat transfer from the converter to the sink being neglected in the model).

Knowing the expression of T_H as a function of ΔS, one gets:

$$W = \left(T_{HSi} \frac{G_H}{G_H + \Delta S} - T_0 \right) \Delta S \left(1 - T_0 \frac{d_{IS}}{T_{HSi} - T_0} \right). \tag{59}$$

The optimum of the mechanical work with respect to ΔS is obtained after some calculations as:

$$MaxW = G_H\left(\sqrt{T_{HSi}} - \sqrt{T_0}\right)^2\left(1 - \frac{T_0}{T_{HSi} - T_0}d_{IS}\right).$$ (60)

The corresponding first law efficiency is:

$$\eta_I(MaxW) = \left(1 - \sqrt{\frac{T_0}{T_{HSi}}}\right)\left(1 - \frac{T_0\,d_{IS}}{T_{HSi} - T_0}\right),$$ (61)

These two quantities are decreasing with the parameter d_{IS}. Thus, the mechanical work W is diminished by the term $\left(1 - T_0\,\frac{d_{I,S}}{T_{HSi} - T_0}\right)$, as shown in (59). Also, (61) clearly proves that irreversibility decreases the efficiency corresponding to maximum work.

4. Conclusions

This work presents a short history of the Carnot cycle approaches developed in the frame of Equilibrium Thermodynamics, of the actual endo-irreversible Carnot engine in Irreversible Thermodynamics, and for the Chambadal model.

The analytical model developed in the frame of Equilibrium Thermodynamics integrates the irreversibility in the natural form of an entropy production, or in an entropic ratio form (with various justified and correlated definitions) and is mainly related to the irreversibility degree introduced by Novikov [32].

The mechanical work of the irreversible cycle remains linked to the form of the irreversible isotherms and adiabatics, as has been shown in Section 2.1. The irreversibilities distribution between isothermal and adiabatic processes leads to framing the actual work between zero and max $MaxW$, with the corresponding efficiencies pertaining to the interval between zero and the Carnot efficiency.

For the Chambadal model, the cycle approach or the one in the stationary dynamic regime come down to the same form, by introducing a characteristic property G_H, which is representative of the finite physical dimensions connected to the reference heat transfer entropy ΔS, characterizing the converter.

It results in the existence of an optimum of the mechanical work either in terms of intensive variable T_H, or in terms of extensive variable ΔS.

The two approaches leading to coherent results, but with different interpretations, provide a new perspective on the optimization of the Carnot engine, mainly by coupling of the hot source with the converter (G_H and ΔS).

Extensions are in progress, which are based on more complete models, particularly on the Curzon-Ahlborn one, that considers the non-equilibrium at source and sink. Also, the stochastic thermodynamics approach [33,34] is intended to be used in future works to evaluate different sources of irreversibility in the corresponding heat engines exactly, without additional assumptions. It could offer an alternative to evaluate the analytical results provided by our phenomenological approach.

Author Contributions: M.F. contributed to the development of the model, M.C. contributed to the preparation of the manuscript, design of the study and final proof, and to the analysis and interpretation of the results. All authors have read and agreed to the published version of the manuscript.

Funding: This research received no external funding.

Conflicts of Interest: The authors declare no conflict of interest.

Symbols

A	heat transfer area, [m^2]
c_p	specific heat at constant pressure, [J/(kg K)]
$\dot{C}$	heat capacity rate, [W/K]
d	Novikov's irreversibility degree, [-]
$ECOP$	ecological coefficient of performance, [-]
G	finite physical dimension, [J/K]
K	heat transfer conductance, [W/K]
I	entropic ratio, [-]
NTU	number of heat transfer units, [-]
Q	heat, [J]
$\dot{Q}$	heat transfer rate, [W]
S	entropy, [J/K]
T	temperature, [K]
U	overall heat transfer coefficient, [W/(m^2 K)]
W	mechanical work, [J]

Greek Symbols

ε	effectiveness, [-]
η	efficiency, [-]

Subscripts

C	relative to the cycled fluid at the cold source
Co	relative to compression
CS	cold source or sink
Ex	relative to expansion
H	relative to the cycled fluid at the hot source
HS	hot source
HSi	hot source
I	irreversibility of the converter
IS	irreversibility of the system (source-converter-sink)
S	source or sink, relative to temperature difference of source and sink
T	total, relative to extreme temperature difference of the cycled fluid
0	ambient

References

1. Carnot, S. *Réflexion sur la Puissance Motrice du feu*; Bachelier: Paris, France, 1824. (In French)
2. Carnot, S. *Principes Fondamentaux de L'équilibre et du Mouvement*; Wentworth Press: Paris, France, 2018. (In French)
3. Andresen, B.; Berry, R.S.; Ondrechen, M.I. Thermodynamics for processes in finite time. *Acc. Chem. Res.* **1984**, *17*, 266–271. [CrossRef]
4. Ibrahim, O.M.; Klein, S.A.; Mitchell, J.W. Optimum heat power cycles for specified boundary conditions. *J. Eng. Gas Turb. Power* **1991**, *113*, 514–521.
5. Feidt, M. *Thermodynamique Optimale en Dimensions Physiques Finies*; Hermes: Paris, France, 2013. (In French)
6. Sánchez Salas, N.; Velasco, S.; Calvo Hernández, A. Unified working regime of irreversible Carnot-like heat engines with nonlinear heat transfer laws. *Energy Convers. Manag.* **2002**, *43*, 2341–2348. [CrossRef]
7. Chen, L.; Xia, S.; Sun, F. Maximum power output of multistage irreversible heat engines under a generalized heat transfer law by using dynamic programming. *Sci. Iran.* **2013**, *20*, 301–312.
8. Feidt, M.; Costea, M.; Petrescu, S.; Stanciu, C. Nonlinear Thermodynamic analysis and optimization of a Carnot engine cycle. *Entropy* **2016**, *18*, 243. [CrossRef]
9. Zhan, Y.-B.; Ma, P.-C.; Zhu, X.-Q. Ecological optimization for a generalized irreversible Carnot engine with an universal heat transfer law. *Procedia Environ. Sci.* **2011**, *11 Pt B*, 945–952. [CrossRef]
10. Stanciu, C.; Feidt, M.; Costea, M.; Stanciu, D. Optimization and entropy production: Application to Carnot-like refrigeration machines. *Entropy* **2018**, *20*, 953. [CrossRef]

11. Park, H.; Kim, M.S. Performance analysis of sequential Carnot cycles with finite heat sources and heat sinks and its application in organic Rankine cycles. *Energy* **2016**, *99*, 1–9. [CrossRef]

12. Feidt, M.; Costea, M.; Petre, C.; Petrescu, S. Optimization of direct Carnot cycle. *Appl. Therm. Eng.* **2007**, *27*, 829–839. [CrossRef]

13. Feidt, M. Optimal Thermodynamics—New upperbounds. *Entropy* **2009**, *11*, 529–547. [CrossRef]

14. Kodal, A.; Sahin, B.; Yilmaz, T. A comparative performance analysis of irreversible Carnot heat engines under maximum power density and maximum power conditions. *Energy Convers. Manag.* **2000**, *41*, 235–248. [CrossRef]

15. Feidt, M. *Thermodynamique et Optimisation Énergétique des Systèmes et Procédés*, 1st ed.; Lavoisier TEC et DOC: Paris, France, 1987. (In French)

16. Vaudrey, A.; Lanzetta, F.; Feidt, M.H.B. Reitlinger and the origin of efficiency at maximum power formula for heat engines. *J. Non-Equil. Thermodyn.* **2014**, *39*, 199–203. [CrossRef]

17. Dong, Y.; El-Bakkali, A.; Feidt, M.; Descombes, G.; Périlhon, C. Association of finite-dimension thermodynamics and bond-graph approach for modeling on irreversible heat engine. *Entropy* **2012**, *14*, 642–653. [CrossRef]

18. Sieniutycz, S.; Jeżowski, J. *Energy Optimization in Process Systems and Fuel Cells*, 3rd ed.; Elsevier: London, UK, 2018.

19. Long, R.; Liu, W. Ecological optimization for general heat engine. *Phys. A Stat. Mech. Appl.* **2015**, *434*, 232–239. [CrossRef]

20. Yu, J.; Zhou, Y.; Liu, Y. Performance optimization of an irreversible Carnot refrigerator with finite mass flow rate. *Int. J. Refrig.* **2011**, *34*, 567–572.

21. Feidt, M.; Le Saos, K.; Costea, M.; Petrescu, S. Optimal allocation of HEX inventory associated with fixed power output or fixed heat transfer rate input. *Int. J. Thermodyn.* **2002**, *5*, 25–36.

22. Feidt, M.; Costea, M. From Finite Time to Finite Dimensions Thermodynamics: The Carnot engine and Onsager's relations revisited. *J. Non-Equil. Thermodyn.* **2018**, *43*, 151–161. [CrossRef]

23. Blaise, M.; Feidt, M.; Maillet, D. Influence of the working fluid properties on optimized power of an irreversible finite dimensions Carnot engine. *Energy Convers. Manag.* **2018**, *163*, 444–456. [CrossRef]

24. Feidt, M. Thermodynamics applied to reverse cycle machines, a review. *Int. J. Refrig.* **2010**, *33*, 1327–1342. [CrossRef]

25. Feidt, M. Thermodynamics of energy systems and processes: A review and perspectives. *J. Appl. Fluid Mech.* **2012**, *5*, 85–98.

26. Han, Y.; Wang, D.; Zhang, C.; Zhu, Y. The entransy degeneration and entransy loss equations for the generalized irreversible Carnot engine system. *Int. J. Heat Mass Transf.* **2017**, *106*, 895–907. [CrossRef]

27. Feidt, M. *Finite Physical Dimensions Optimal Thermodynamics 1—Fundamentals*; ISTE Press—Elsevier: London, UK, 2017; pp. 75–97.

28. Bhattacharyya, S.; Blank, D.A. Power optimized work limit for internally irreversible reciprocating engines. *Int. J. Mech. Sci.* **2000**, *42*, 1357–1368. [CrossRef]

29. Chambadal, P. *Les Centrales Nucléaires*; Armand Colin: Paris, France, 1957. (In French)

30. Curzon, F.L.; Ahlborn, B. Efficiency of a Carnot engine at maximum power output. *Am. J. Phys.* **1975**, *43*, 22–24. [CrossRef]

31. Wang, J.; He, J. Efficiency at maximum power output of an irreversible Carnot-like cycle with internally dissipative friction. *Phys. Rev.* **2012**, *E86*, 051112. [CrossRef] [PubMed]

32. Novikov, I.I. The efficiency of atomic power stations. *At. Energy* **1957**, *3*, 409–412. [CrossRef]

33. Esposito, M.; Kawai, R.; Lindenberg, K.; Van den Broeck, C. Efficiency at maximum power of low dissipation Carnot engines. *Phys. Rev. Lett.* **2010**, *105*, 150603. [CrossRef]

34. Holubec, V.; Ryabov, A. Cycling tames power fluctuations near optimum efficiency. *Phys. Rev. Lett.* **2018**, *121*, 120601. [CrossRef]

Article

OTEC Maximum Net Power Output Using Carnot Cycle and Application to Simplify Heat Exchanger Selection

Kevin Fontaine [1,*], Takeshi Yasunaga [2] and Yasuyuki Ikegami [2]

1   Graduate School of Science and Engineering, Saga University, 1 Honjo-Machi, Saga 840-8502, Japan
2   Institute of Ocean Energy, Saga University, 1 Honjo-Machi, Saga 840-8502, Japan;
    yasunaga@ioes.saga-u.ac.jp (T.Y.); ikegami@cc.saga-u.ac.jp (Y.I.)
*   Correspondence: fontaine.kevin.d@gmail.com

Received: 25 October 2019; Accepted: 18 November 2019; Published: 22 November 2019

Abstract: Ocean thermal energy conversion (OTEC) uses the natural thermal gradient in the sea. It has been investigated to make it competitive with conventional power plants, as it has huge potential and can produce energy steadily throughout the year. This has been done mostly by focusing on improving cycle performances or central elements of OTEC, such as heat exchangers. It is difficult to choose a suitable heat exchanger for OTEC with the separate evaluations of the heat transfer coefficient and pressure drop that are usually found in the literature. Accordingly, this paper presents a method to evaluate heat exchangers for OTEC. On the basis of finite-time thermodynamics, the maximum net power output for different heat exchangers using both heat transfer performance and pressure drop was assessed and compared. This method was successfully applied to three heat exchangers. The most suitable heat exchanger was found to lead to a maximum net power output 158% higher than the output of the least suitable heat exchanger. For a difference of 3.7% in the net power output, a difference of 22% in the Reynolds numbers was found. Therefore, those numbers also play a significant role in the choice of heat exchangers as they affect the pumping power required for seawater flowing. A sensitivity analysis showed that seawater temperature does not affect the choice of heat exchangers, even though the net power output was found to decrease by up to 10% with every temperature difference drop of 1 °C.

Keywords: ocean thermal energy conversion (OTEC); plate heat exchanger; optimization; maximum power output; finite-time thermodynamics

---

1. Introduction

In 2015, the CO_2 emission due to energy generation and heat production was up to 13 540 million tons [1]. Efforts are to be made to lower this figure, especially to meet the Paris Agreement's goal to limit global warming to below 2 °C [2]. Thus, it is necessary to develop renewable energies, which are not represented enough in today's energy mix [3]. A drawback of most implemented renewable energies is their intermittency, therefore, they cannot be used for baseload energy demand without a storage system breakthrough. However, to generate electricity, ocean thermal energy conversion (OTEC) uses the difference between the surface seawater and the deep seawater temperature in tropical areas. As such areas present very low temperature change throughout the year, a steady power generation can be achieved. Moreover, OTEC has huge potential, as its resources are estimated at a maximum of 7 TW of net energy production [4]. In addition to power generation, it is possible with such a system to produce freshwater using the warm seawater. This water can be used to create pure hydrogen to store or transport the energy generated by the power plant. In addition to tackling climate change and providing clean energy, OTEC can contribute to reaching five other sustainable development goals defined in 2015 by the general assembly of the United Nations [5]. Indeed, OTEC produces deep seawater as a byproduct, which can be used for aquaculture and desalination. The implementation of such a power plant could play a major role in the economic growth of cities and countries, especially on islands. It would also promote industry, innovation, infrastructure, as well as providing employment for the construction and operation of the plant, auxiliaries, and other industries that make use of deep seawater.

As it uses the low-temperature gradient in the ocean, which is 20–25 °C in suitable locations, the OTEC system presents a very low maximal theoretical thermal efficiency of 3%–5%. However, the system is a potential substitute for conventional power plants with higher efficiency, considering that the resource has virtually no cost. It is necessary to understand the basic theory, known as finite-time thermodynamics (FTT), which was first applied to OTEC by Wu [6] and extended by Ikegami and Bejan [7]. They constructed the basic theory for OTEC using FTT; however, the application in engineering is still limited. It is also necessary to optimize, as much as possible, the net power output of an OTEC power plant. This can be done in different ways: by optimizing cycles or investigating more efficient ones, by choosing the most suitable working fluid, or by investigating key elements of the system.

To harvest ocean thermal energy, it is possible to use a flash chamber to evaporate the warm surface seawater. The steam is used to operate a turbine and is then condensed by the cold deep seawater. In doing so, both energy and freshwater are produced. Kim et al. researched such a system, namely, open-cycle OTEC [8]. They thoroughly investigated the system using three different condensers in order to adjust the power generation and desalination. They found that there is an optimal fraction of steam that enters the turbine that maximizes the ratio of the generated power over the cold water flow rate, therefore optimizing the use of the said steam throughout the plant. Kim et al. also investigated the possible replacement of the flash evaporation chamber by a vacuum membrane distillation module, which would require less than 10% of the volume normally occupied by the evaporator and result in a decrease of the open-cycle OTEC plant size by 30% and the electricity cost by 2.1% [9]. In contrast to the open cycle, the closed cycle uses a working fluid in a closed-loop, and is the main focus of this study, as it is more suitable to generate large amount of power. Hereafter, only the closed cycle is discussed.

Regarding cycle investigations, D.H. Johnson calculated the exergy available in a specific amount of both cold and warm water and used the results to compare the performance of different OTEC cycles, such as the Beck and Rankine cycles. He used the second law efficiency of the cycles, defined as the ratio between the cycle power output and the exergy from ocean thermal resources [10]. His study concluded the triple Rankine cycle as the one which presents the most potential. He also compared the second law efficiency of a coal-fired Rankine cycle with a single-stage OTEC Rankine cycle and found that "an OTEC plant uses the exergy of the ocean thermal resource about as efficiently as a conventional coal-fired plant uses the exergy of coal". In 1982, Kalina presented a cycle that introduced the regeneration of the working fluid [11]. This cycle has been widely investigated: Zhang et al. made a review of this research and realized a comparison with the Rankine cycle [12]. They found that, in most cases, the Kalina cycle presents better efficiencies than the Rankine cycle, especially in the case of low-temperature resources. Uehara et al. then proposed a new cycle, as an improvement of the Kalina cycle, and implemented a second turbine to the cycle [13]. The study showed an increase of ~10% in the thermal efficiency compared to a Kalina cycle and more than 30% compared to a single-stage Rankine cycle for specific conditions. Ikegami et al. investigated the reduction of irreversible losses in the heat exchange process for a double-stage Rankine cycle [14]. This way, they demonstrated that an OTEC cycle net power output could be increased using this cycle or the Kalina cycle. The current study initially focused on the Carnot cycle, as it is the basic theoretical cycle for OTEC. After that, a Rankine cycle was investigated, as it is a more realistic cycle and for comparison purposes. Additionally, calculations from the Rankine cycle can be used as a base for multistage Rankine cycle calculations in the future, as they have been found to have great potential.

For fluid choices, Dijoux et al. built a model of an organic Rankine cycle applied to OTEC to compare the performances of 26 different working fluids. They managed to select three fluids that would be suitable for OTEC by considering several parameters, including thermodynamics performances, even though no fluid could meet all the ideal parameters [15]. One of the selected fluids was ammonia, which has also been identified as the fluid of choice by Bernardoni et al. [16]. Thus, this study considered ammonia when working fluid properties were required.

In OTEC, heat exchangers allow heat harvest and are pointed out by many studies as one of the most important parts of the system to investigate. Sinama et al. performed an OTEC cycle optimization by minimizing the destroyed exergy [17]. The study presents a sensitivity analysis that shows the role of heat exchangers in an OTEC cycle as well as the importance of optimizing them or developing more suitable ones. Sun et al. reached the same results in their optimization design and exergy analysis of an organic Rankine cycle [18]. Bernardoni et al. realized a techno-economic analysis of a closed OTEC cycle to assess its levelized cost of energy (LCOE) [16]. This is defined by the sum of the operating cost of a power plant over its lifetime and its capital cost divided by the net energy production over its lifetime. Bernardoni et al. identified heat exchangers as one of the most expensive constituents of an OTEC power plant, making them suitable candidates for further study. Accordingly, this paper focused on investigating heat exchangers for OTEC.

Many types of heat exchangers exist, and the most investigated one in OTEC is the plate heat exchanger. Indeed, the compactness of this type of heat exchanger compared to other types allows reaching the significant heat transfer area required for OTEC purposes with less material. It is necessary to compare the maximum net power output of a power plant for different heat exchangers to choose the most suitable one. For this purpose, it is first necessary to evaluate the net power output of an OTEC system for a given heat exchanger, which depends not only on the heat transfer performance of the heat exchanger but also its pressure drop. Due to the low thermal efficiency of the OTEC cycle, the pumping power required to counter the pressure drop that occurs inside the heat exchanger is of the same order of magnitude as the gross power generated by the harvested heat. Yasunaga et al. [19] realized an OTEC performance evaluation in terms of FTT based on a Carnot cycle and identified a theoretical relationship between heat transfer performance, pressure drop, and OTEC net power output. They also showed that a compromise must be found, as both parameters are not independent. To evaluate the net power output of an OTEC power plant, it is thus necessary to know both the heat transfer performance and the pressure drop of the heat exchanger. Although they are independently evaluated, both these parameters have been widely investigated [20–22]. Uehara and Ikegami conducted an in-depth optimization of a closed-cycle OTEC system with all its constituents for different temperatures of the warm seawater [23]. They used the ratio of the heat transfer area over the cycle net power output as the objective function to minimize the cost of the produced electricity, which greatly depends on the size of the heat exchangers. In doing so, they could calculate, among other things, the optimized, required heat transfer area for a specific evaporator and condenser set. However, studies that have evaluated an OTEC system net power output as a function of both the heat transfer performance and the pressure drop, to allow a comparison between different heat exchangers, are scarce. The method used by Uehara et al., although very accurate and quite exhaustive, is not suitable for heat exchanger comparisons due to its high complexity. It is therefore necessary to develop a relatively easier method that will allow such a comparison. This was the aim of the present study, and the main focus was on the trade-off between the pressure drop and the heat transfer performance; losses occurring on other components should not affect the comparison results.

The goal of this work was to find the most suitable heat exchanger for OTEC, which is useful when designing a new power plant. A simplified method was developed to evaluate and then maximize the net power output of an OTEC system for a given heat exchanger. As it is one of the most expensive components, it is necessary to select the heat exchanger that leads to the highest net power output for a minimal amount of material [16–18]. This study, therefore, focused on the net power output per unit of the heat exchanger surface area. Increasing the heat transfer area on plate heat exchangers, the preferred type of heat exchanger for OTEC, can be easily achieved by adding more plates. This will not alter the results as long as the Reynolds numbers inside a plate and the number of passes remains the same. The results for three different heat exchangers from the literature were compared. First, calculations were done for the Carnot cycle. Then, Rankine cycle calculations were realized to investigate the differences between the two cycles. The influence of seawater temperature was investigated through a sensitivity analysis. Indeed, each OTEC power plant is designed to operate at specific conditions, and the heat source temperature affects their performance. This method will not be able to predict the net power output and optimum operating points as accurately as what was achieved by Uehara and Ikegami [23]; however, its relative ease of use will make it preferable for heat exchanger comparisons. In addition, it may constitute the basis for less complex methods than the one proposed by Uehara and Ikegami, but with comparable accuracy. The proposed method presents the following:

- It addresses the lack of evaluation methods to efficiently select a heat exchanger for OTEC purposes and considers the trade-off between heat transfer performance and pressure drop.
- It is easily applicable for different heat exchangers as long as geometry, heat transfer coefficient correlation, and pressure drop correlation are provided.
- It is easily applicable to different seawater temperatures.

2. Description and Analysis

The first step before the actual optimization is to define the objective function that will be maximized. In this study, the comparison of the heat exchangers was based on the net power output of the power plant per unit of the heat exchanger surface area. The net power output of an OTEC system depends on the gross power output that can be produced by the harvested heat and the different losses that occur in the system.

In OTEC, the heat from the sea is harvested using plate heat exchangers. A plate heat exchanger consists of a stack of plates in which the thermal energy of a relatively hot fluid is transferred to a colder one, with each fluid flowing on a different side of a plate. Therefore, the harvested heat depends on the convection heat transfer coefficient of both fluids and the heat transfer conductivity of the plate through which the heat transfer occurs. In addition, deposit layers of different natures are observed on the heat exchanger plate as the OTEC plant operates. This is called fouling and is caused by elements; organic matter; or living organisms present in the seawater. Such layers add thermal resistance to the plate, and thus decrease the heat transfer performance of the exchanger.

Part of the gross power output generated by the power plant is used for water ducting. In addition, losses happen in the heat transfer process itself and in the form of pressure drop that will be countered using pumps, decreasing the net power output of the system. The pressure drop happens in the pipes in which each fluid circulates and in the heat exchangers.

In this study, an optimization was realized for an OTEC system based on a Carnot and a Rankine cycles. For both cycles, the following assumptions were considered:

- The heat transfer coefficient of the working fluid is much greater than the seawater one as the working fluid undergoes a phase change [24].
- The thermal resistance due to fouling can be neglected.
- The pressure drop on the working fluid side can be neglected [16,17,23].
- Changes in the water thermodynamic properties in the heat exchangers due to temperature variation can be neglected.

Moreover, although it has a non-negligible impact on the net power output of an OTEC power plant, the pressure drop that occurs in the piping system is not accounted for in the present paper. Indeed, it mainly depends on the length of the pipes and therefore, it has no impact on the choice of a heat exchanger.

2.1. Carnot Cycle: Concept and Equations

In this section, the conceptual OTEC system in Figure 1a is assumed to be working on a Carnot cycle, which is described in Figure 1b. A working fluid is heated by the warm seawater in a first heat exchanger in which it evaporates. The resulting vapor is then used to operate a turbine before being condensed by the deep seawater through a second heat exchanger. Finally, the working fluid is pumped into the evaporator for another cycle. This is an ideal cycle using isothermal heat exchange as well as isentropic compression and expansion processes, although they cannot be achieved in a practical power plant.

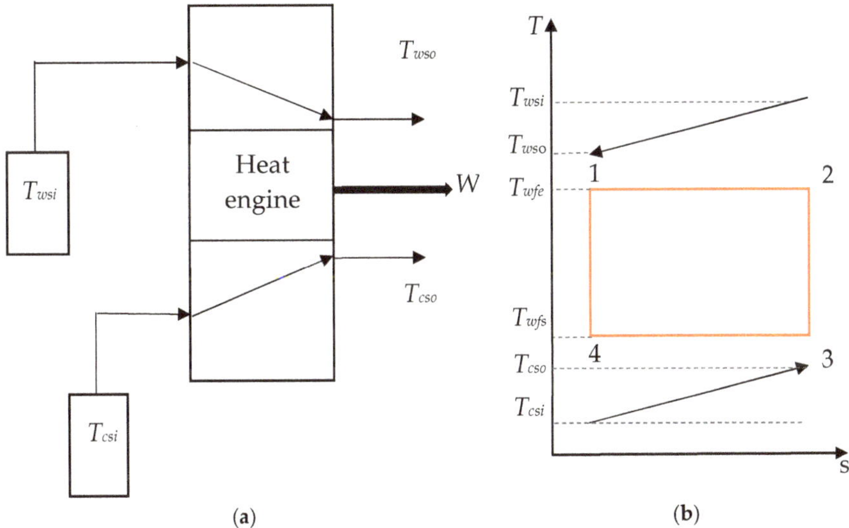

Figure 1. (**a**) Model of the ocean thermal energy conversion (OTEC) system and (**b**) temperature–entropy diagram of a Carnot cycle.

As this study focuses on the cycle performance, the pumping power required for water ducting is not considered. Thus, the net power output is equal to the difference between the gross power generated by the heat engine, W_{gross}, and the required pumping power to counter the seawater pressure drop in the heat exchanger, W_P. The net power output, W_{net}, of the Carnot cycle is therefore given by Equation (1) [7]:

$$W_{net} = W_{gross} - W_P \tag{1}$$

The power generated by the cycle is given by the balance between the heat added to the system (Q_e) and the heat taken from the system (Q_c) Equation (2) [25]:

$$\begin{aligned}
W_{gross} = Q_e - Q_c &= C_{ws}\epsilon_{ws}\left(T_{wsi} - T_{wfe}\right) - C_{cs}\epsilon_{cs}\left(T_{wfc} - T_{csi}\right) \\
&= (mc_p\epsilon)_{ws}\left(T_{wsi} - T_{wfe}\right) - (mc_p\epsilon)_{cs}\left(T_{wfc} - T_{csi}\right)
\end{aligned} \tag{2}$$

where ε is the heat exchanger efficiency, which is equal to the ratio of the actual heat exchange that occurs in the heat exchanger and the maximum theoretical heat exchange that could occur for an infinite counterflow heat exchanger without losses. C is the product of the mass flow rate, m, and the specific heat capacity, c_p, of the seawater. Subscript ws stands for warm source and cs stands for cold source. T_{wsi}, T_{csi}, T_{wfe}, and T_{wfc} are the temperature of the warm seawater at the inlet of the evaporator, the temperature of the cold seawater at the inlet of the condenser, the temperature of the working fluid in the evaporator, and the temperature of the working fluid in the condenser, respectively.

In the case of an OTEC power plant, as a phase change occurs in both heat exchangers, the efficiency, ε, can be written as Equations (3) and (4) [25]:

$$\epsilon_{ws} = 1 - \exp(-\mathrm{NTU}_{ws}) \tag{3}$$

$$\epsilon_{cs} = 1 - \exp(-\mathrm{NTU}_{cs}) \tag{4}$$

where NTU is the number of transfer units and is defined in Equation (5):

$$\mathrm{NTU} = \frac{UA}{C} \tag{5}$$

where U is the overall heat transfer coefficient and A is the heat transfer area of the heat exchanger.

The Lagrange multiplier method is applied, similar to Ibrahim et al., to find the conditions on the temperatures that maximize the power output of the heat engine, as can be seen from Equations (6) to (9) [26]. The entropy balance written in Equation (6) is used as the constraint function, as it is equal to 0 when there are no irreversibilities:

$$\Delta s = \frac{C_{ws}\epsilon_{ws}\left(T_{wsi} - T_{wfe}\right)}{T_{wfe}} - \frac{C_{cs}\epsilon_{cs}\left(T_{wfc} - T_{csi}\right)}{T_{wfc}} = 0 \tag{6}$$

The Lagrange multiplier, Λ, is then introduced and verifies Equation (7):

$$\frac{\delta W_{gross}}{\delta T_{wfe}} = \Lambda \frac{\delta \Delta s}{\delta T_{wfe}} \quad \text{and} \quad \frac{\delta W_{gross}}{\delta T_{wfc}} = \Lambda \frac{\delta \Delta s}{\delta T_{wfc}} \tag{7}$$

It leads, using Equations (2), (6), and (7), to the following relationship at the maximum point.

$$\frac{T_{wfc}}{T_{wfe}} = \sqrt{\frac{T_{csi}}{T_{wsi}}} \tag{8}$$

From here, it is possible to express the maximum power output of the heat engine, as in [7,19,26]:

$$W_{gross} = \frac{\left(\sqrt{T_{wsi}} - \sqrt{T_{csi}}\right)^2}{\frac{1}{C_{ws}\epsilon_{ws}} + \frac{1}{C_{cs}\epsilon_{cs}}} \tag{9}$$

In the case of OTEC, as heat exchangers are the most expensive components [16], the goal of optimization is to maximize the net power output per unit of heat exchanger surface area. Therefore, the objective function is defined as the net power output of the heat engine, assuming it operates at the optimized temperature ratio shown in Equation (8) [7], divided by the total surface area of both the evaporator and condenser. In addition, heat exchangers, for which the calculation is made, do not have the same surface area. Thus, given that the surface area can be changed by the addition or subtraction of plates, the net power output per unit of surface area, w_{net}, allows a more efficient comparison. It is expressed in Equation (10), which is deduced from Equations (1) and (9):

$$w_{net} = \frac{1}{\left(As_{ws} + As_{cs}\right)} \left(\frac{\left(\sqrt{T_{wsi}} - \sqrt{T_{csi}}\right)^2}{\frac{1}{C_{ws}\left(1 - e^{-\mathrm{NTU}_{WS}}\right)} + \frac{1}{C_{cs}\left(1 - e^{-\mathrm{NTU}_{WS}}\right)}} - W_{Pws} - W_{Pcs} \right) \tag{10}$$

where As is the heat exchanger surface area calculated as $As = LWi(number\ of\ plates - 2)$, with L and Wi representing the length and width of the plate, respectively. As differs from A as it does not take into account any plate surface patterns, such as herringbone. W_P is defined as in [7]:

$$W_P = \frac{2fLRe^3\mu_w^3 S}{D^4\rho_w^2}$$ (11)

where D is the equivalent diameter of the heat exchanger, L is the length of a plate and the Reynolds number, Re, is defined as:

$$Re = \frac{\rho_w v D}{\mu_w}$$ (12)

with ρ_w the seawater density, v the seawater mean velocity, and μ_w the seawater dynamic viscosity. f is the friction factor and is defined as:

$$f = \frac{\tau}{\frac{\rho_w v^2}{2}}$$ (13)

where τ is the wall shear stress of the heat exchanger. However, in this study, an experimental correlation of the friction factor is used. It is assumed that the friction factor can be written as:

$$f = \beta Re^\xi$$ (14)

where β and ξ are two constants that depend on the heat exchanger properties.

UA, from Equation (5), is defined as follows:

$$\frac{1}{UA} = \frac{1}{\alpha_w A} + \frac{t}{\lambda_p A} + \frac{1}{\alpha_{wf} A} + \frac{R_f}{A}$$ (15)

where α_w is the seawater convective heat transfer coefficient, α_{wf} is the working fluid heat transfer coefficient, t is the plate thickness, R_f is the resistance due to fouling, and λ_p is the thermal conductivity of the plate.

As specified at the beginning of Section 2, fouling is neglected and α_{wf} is assumed to be much greater than α_w. Therefore:

$$\frac{1}{\alpha_w A} + \frac{t}{\lambda_p A} + \frac{1}{\alpha_{wf} A} + \frac{R_f}{A} \approx \frac{1}{\alpha_w A} + \frac{t}{\lambda_p A} = \frac{1}{\alpha_w A} + B$$ (16)

where B is a constant and α_w is calculated from the Nusselt number (Nu), defined as follows, with λ_w being the water thermal conductivity and D the equivalent diameter:

$$Nu = \frac{\alpha_w D}{\lambda_w}.$$ (17)

Furthermore, the following assumption is taken:

$$Nu = dRe^\gamma Pr^n,$$ (18)

with d, γ, and n being constant coefficients for the Nusselt correlation. The Prandtl number, Pr, is defined as:

$$Pr = \frac{\mu_w c_p}{\lambda_w},$$ (19)

leading to:

$$UA = \frac{ANu\lambda_w}{D + BANu\lambda_w} = \frac{AdRe^\gamma Pr^n \lambda_w}{D + BAdRe^\gamma Pr^n \lambda_w}$$ (20)

In addition, using Prandtl and Reynolds numbers definitions from Equations (12) and (19), C can be written as:

$$C = mC_p = \rho v S C_p = \frac{RePr\lambda_w S}{D}$$ (21)

Therefore,

$$\text{NTU} = \frac{UA}{C} = \frac{D}{\text{RePr}\lambda_w S}\frac{AdRe^{\gamma}\text{Pr}^n\lambda_w}{D+BAdRe^{\gamma}\text{Pr}^n\lambda_w} = \frac{dRe^{\gamma-1}\text{Pr}^{n-1}AD}{(D+BdRe^{\gamma}\text{Pr}^n\lambda_w A)S} \tag{22}$$

Then, replacing NTU and C in Equation (10) by their respective expressions described in Equations (22) and (21), the following equation of the net power output per unit of heat exchanger surface area, w_{net}, is deduced:

$$w_{net} = \frac{1}{As_{ws}+As_{cs}}\frac{\left(\sqrt{T_{ws}}-\sqrt{T_{CS}}\right)^2}{\left(\frac{\text{RePr}\lambda S}{D}\left(1-exp\left(-\frac{dRe^{\gamma-1}\text{Pr}^{n-1}AD}{(D+BdRe^{\gamma}\text{Pr}^n\lambda A)S}\right)\right)\right)_{WS}}+\frac{1}{\left(\frac{\text{RePr}\lambda S}{D}\left(1-exp\left(-\frac{dRe^{\gamma-1}\text{Pr}^{n-1}AD}{(D+BdRe^{\gamma}\text{Pr}^n\lambda A)S}\right)\right)\right)_{CS}}}-\frac{1}{As_{ws}+As_{cs}}\left(\left(\frac{2fLRe^3\mu^3 S}{D^4\rho^2}\right)_{WS}-\left(\frac{2fLRe^3\mu^3 S}{D^4\rho^2}\right)_{CS}\right) \tag{23}$$

It has been possible here to express the net power output of the heat engine as a function of the properties of the seawater side only, which will facilitate the calculations.

As the Carnot cycle is ideal, it does not reflect the actual net power output of an OTEC system. It is possible to introduce a FTT irreversibility factor, as done by Ibrahim et al. [26]; however, this coefficient regroups different sources of internal irreversibilities, which makes it difficult to assess, and is therefore not usable for practical applications like the OTEC power plant. A better way is to proceed with a cycle that can be used for OTEC, such as the Rankine cycle. A comparison must be carried out to see if the use of a different cycle will have an impact on the choice of a heat exchanger.

2.2. Rankine Cycle: Concept and Equations

For the Rankine cycle, described in Figure 2a,b, the net power output can be given by Equation (24). It was assumed that the working fluid at the outlet of the heat exchangers was in a saturated state. This cycle differs from the Carnot cycle, as it can be used practically in power plants. In this case, heat exchange is not isothermal, and a change in entropy occurs in both compression and expansion processes.

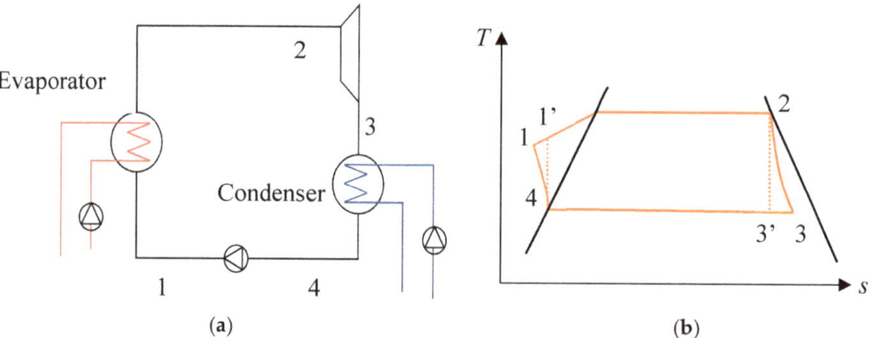

Figure 2. (a) Description of the Rankine cycle and (b) temperature–entropy diagram of a Carnot cycle.

As for the Carnot cycle, the objective function is the net power output of the heat engine divided by the total heat exchanger surface area:

$$w_{net} = \frac{1}{As_{ws}+As_{cs}}\left(m_{wf}(h_2-h_3-h_1+h_4)-\left(\frac{2\beta Re^{3+\xi}L\mu^3 S}{D^4\rho^2}\right)_{ws}-\left(\frac{2\beta Re^{3+\xi}L\mu^3 S}{D^4\rho^2}\right)_{cs}\right) \tag{24}$$

where m_{wf} is the mass flow rate of the working fluid. h_1, h_2, h_3, and h_4 are the enthalpy values of the corresponding points in Figure 2 and with:

$$h_3 = h_2(1 - \eta_T) + \eta_T h_3' \text{ and } h_1 = h_4(1 - \eta_p) + \eta_p h_1' \tag{25}$$

h'_1 and h'_3 are the enthalpy values of the corresponding points in the case of an isentropic process, and η_T and η_p are the turbine and pump efficiencies, respectively. For the calculations, the turbine efficiency is equal to 0.85 and the pump efficiency is equal to 0.8.

Enthalpy values can be calculated from temperatures T_2 and T_4 at Points 2 and 4 using the software REFPROP [27]. T_2 and T_4 can be evaluated from Equation (26):

$$T_2 = \frac{T_{wsi} - T_{wso}e^{\text{NTU}_{ws}}}{\epsilon_{ws}} \text{ and } T_4 = \frac{T_{csi} - T_{cso}e^{\text{NTU}_{cs}}}{\epsilon_{cs}} \tag{26}$$

h_2 and h_4 are calculated using the respective temperature and because the fluid is at a saturated state. It is possible to calculate h_3 using $T'_3 = T_4$, $s_2 = s'_3$, and Equation (26). For Point 1, $s_4 = s'_1$ and $P_2 = P'_1$; P is the pressure of the fluid and s is its specific entropy. NTU is calculated as detailed for the Carnot cycle in Equations (15)–(22). Details for the calculations to obtain Equation (26) are given in Appendix A.

3. Optimization Process

The optimization was realized for different plate heat exchangers tested by Kushibe et al. [28]. The specifications are given in Table 1; they are all versatile plates. Plate heat exchanger (PHE) 1 is meant for high pressure and high temperature and is used as an evaporator in power generation systems using hot springs as a heat source, PHE 2 is a typical herringbone plate heat exchanger, and PHE 3 was specially invented by Prof. Uehara as both an evaporator and a condenser [28].

Table 1. Heat exchanger specifications. PHE: Plate heat exchanger.

Heat Exchanger	PHE 1	PHE 2	PHE 3
Length L (mm)	960	718	1765
Width Wi (mm)	576	325	605
Thickness t (mm)	0.7	0.5	0.6
Space between plates δ (mm)	4.00	3.95	2.68
Equivalent diameter D (mm)	8.00	7.90	5.36
Material	SUS316	Titanium	Titanium
Thermal conductivity λ_p (W/(m·K))	16.3	21	21
Pattern	Herringbone (72°)	Herringbone (30°)	Fluting and drainage
Number of plates	120	20	52
Total heat transfer area A (m^2)	100.3	3.96	40.6
Total cross surface area S (m^2)	0.140	0.012	0.041

The optimization was run using the "minimize" function written by Rody Oldenhuis based on the derivative-free method (Nelder–Mead simplex algorithm) fminsearch function of Matlab R2017a (MathWorks, Natick, MA, USA) [29]. It searches for a local minimum of a given objective function with several variables. It allows applying boundaries on the variables as well as linear and nonlinear equality and inequality constraints that the objective function needs to satisfy. In the present work, boundaries were specified for all the variables and calculations were performed several times from different starting points generated randomly within the boundaries so that a global minimum could be found. As the goal was to maximize w_{net}, the objective function used with "minimize" was $-w_{net}$.

Calculations were done using the friction factor and Nusselt numbers correlations of the corresponding heat exchanger, as defined in Equations (14) and (18). Table 2 gives the different coefficients for the correlations, which were approximated from linear interpolation from the data of the study realized by Kushibe et al. [28].

Table 2. Coefficients for the Nusselt numbers and friction factor correlations.

Heat Exchanger	d	γ	n	β	ξ
PHE 1	0.111	0.8	1/3	1.4863	−0.0540
PHE 2	0.058	0.8	1/3	6.5059	−0.3292
PHE 3	0.051	0.8	1/3	0.7371	−0.1274

For both cycles, the water properties are taken at pressures of 130 kPa and 170 kPa for warm and cold seawater sources, respectively, for the corresponding temperatures and were obtained using REFPROP [27]. Warm and cold water source temperatures are first set at 30 °C and 5°, respectively. Then, a sensitivity analysis is performed for both cycles with a temperature difference between 23 °C and 20 °C. These temperature differences are obtained by decreasing the warm and cold water source alternatively.

3.1. Carnot Cycle

In the case of the Carnot cycle, the objective function from Equation (23) presents only two variables: the two Reynolds numbers. They were limited to a corresponding water mean velocity between 0.2 m/s and 1.8 m/s, which is the speed range reachable in plate heat exchangers. No constraints were applied to the Carnot cycle.

3.2. Rankine Cycle

In the case of the Rankine cycle, the objective function is given in Equation (24). As for the Carnot cycle, both Reynolds numbers are variables of the objective function. In addition, the mass flow rate of the working fluid, here ammonia is also one of the variables. Finally, water temperatures at the outlet of the heat exchangers are needed to calculate the enthalpies required for the objective function as shown in Equations (24)–(26). These temperatures were therefore also set as variables during the optimization process. This led to a total of five variables: Re_{ws}, Re_{cs}, T_{wso}, T_{cso}, and m_{wf}. As for the Carnot cycle, boundaries were used on the Reynolds number to restrain the water mean velocity between the 0.2 to 1.8 m/s range. Temperatures were taken to be between T_{csi} and T_{wsi}. m_{wf} was restrained to be between 0.001 kg/s and a value that is equal to a water mean velocity of 1.8 m/s, as water mass flow rate is higher than ammonia mass flow rate in OTEC system.

As the optimization is based on enthalpies values, the following constraints were added to ensure that the heat balance (27) and energy conservation (28,29) are respected.

$$C_{ws}(T_{wsi} - T_{wso}) - C_{cs}(T_{cso} - T_{csi}) = m_f(h_2 - h_3 - h_1 + h_4) \tag{27}$$

$$C_{ws}(T_{wsi} - T_{wso}) = m_f(h_2 - h_1) \tag{28}$$

$$C_{cs}(T_{cso} - T_{csi}) = m_f(h_3 - h_4) \tag{29}$$

4. Optimization Results

4.1. Carnot Cycle Results

Figure 3 shows w_{net} as a function of the Reynolds number for PHE 1, PHE 2, and PHE 3. Table 3 shows the heat transfer coefficient and pressure drop of the three heat exchangers.

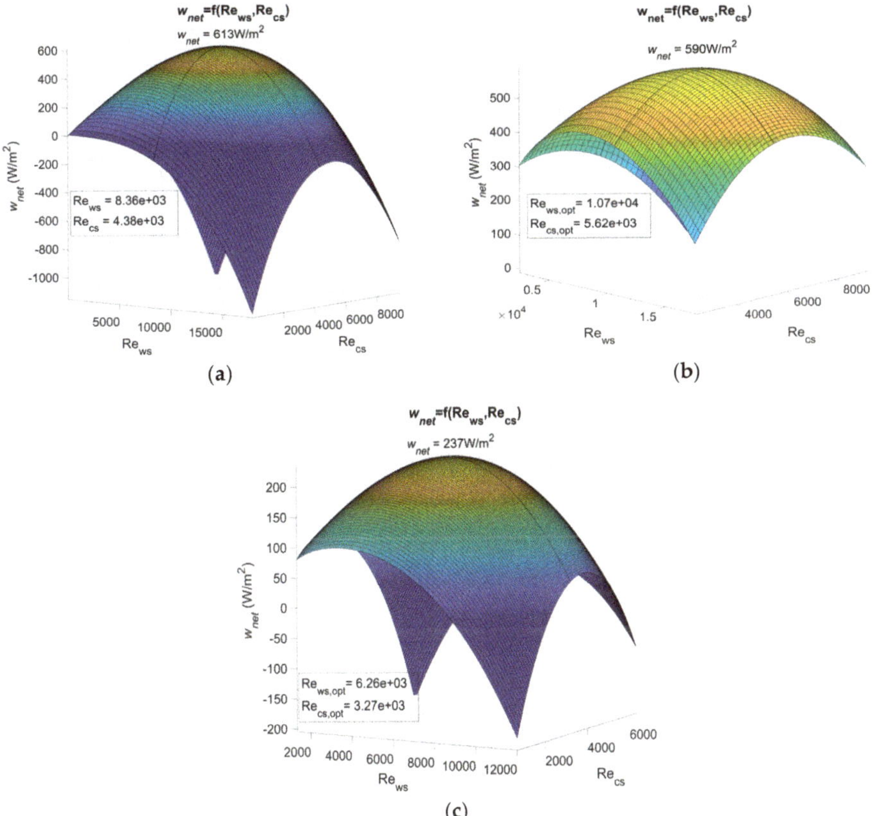

Figure 3. (**a**) Maximum net power output of an OTEC power plant using Plate heat exchanger (PHE) 1 as both evaporator and condenser as a function of Reynolds, (**b**) maximum net power output of an OTEC power plant using PHE 2 as both evaporator and condenser as a function of Reynolds, and (**c**) maximum net power output of an OTEC power plant using PHE 3 as both evaporator and condenser as a function of Reynolds.

Table 3. Heat transfer coefficient and friction factor of the seawater inside the heat exchangers.

Heat Exchanger	α_{ws} (W/m²·K)	α_{cs} (W/m²·K)	f_{ws} (−)	f_{cs} (−)
PHE 1	20 569	14 450	0.913	0.945
PHE 2	13 259	9 334	0.307	0.379
PHE 3	11 191	7 843	0.242	0.263

The actual power output of an OTEC power plant using the specified heat exchanger should be less than these results as calculations are based on a Carnot cycle, and if the pressure drop is considered, the required power for water ducting is not. However, the optimum Reynolds numbers $Re_{ws,opt}$ and $Re_{cs,opt}$ found in these results show that, for a non-negligible range of Reynolds numbers, the net power output can be null or negative (dark blue areas on Figure 3). The flow rate should then be controlled rigorously to adjust the Reynolds numbers in the heat exchangers.

This work allows the performances of heat exchangers to be compared in terms of the OTEC system's maximum net power output per unit of heat exchanger surface area, $w_{net,max}$. These results show the importance of correctly selecting a heat exchanger, as $w_{net,max}$ is highly dependent on which one is used. Indeed, a huge difference was noticed between the results using PHE 3 and the results using the other two at their optimal operating point. PHE 1 led to a $w_{net,max}$ that was 158% higher than the one of PHE 3. As for PHE 2, the $w_{net,max}$ was 149% higher than the one achieved with PHE 3. This difference can be explained by the low heat transfer coefficient of PHE 3.

PHE 2 led to a $w_{net,max}$ that was only 3.7% lower than PHE 1, even though its heat transfer coefficient was 35% lower than the one of PHE 1. This can be explained by a friction factor that was 6 to 66% lower in the case of PHE 2. In addition, PHE 2 was found to be less sensitive to a change in the Reynolds numbers. PHE 1, however, presented optimum Reynolds numbers that were 22% lower than those of PHE 2. From these observations, one can easily conceive a heat exchanger that would lead to a higher $w_{net,max}$ than PHE 1 and also present lower heat transfer coefficient if the pressure drop was low enough. This would be possible because a low pressure drop allows the use of higher Reynolds numbers to compensate for a low heat transfer coefficient. Such a heat exchanger, because of its low pressure drop, would be less affected by a change in the operating Reynolds numbers. A high pressure drop heat exchanger, however, would require relatively lower optimum Reynolds numbers, which implies a lower pumping power for water ducting and/or lower diameter pipes.

The same calculations were carried out for different hot and cold seawater temperatures to figure out how they would affect the net power output for each heat exchanger. The results are given in Figure 4.

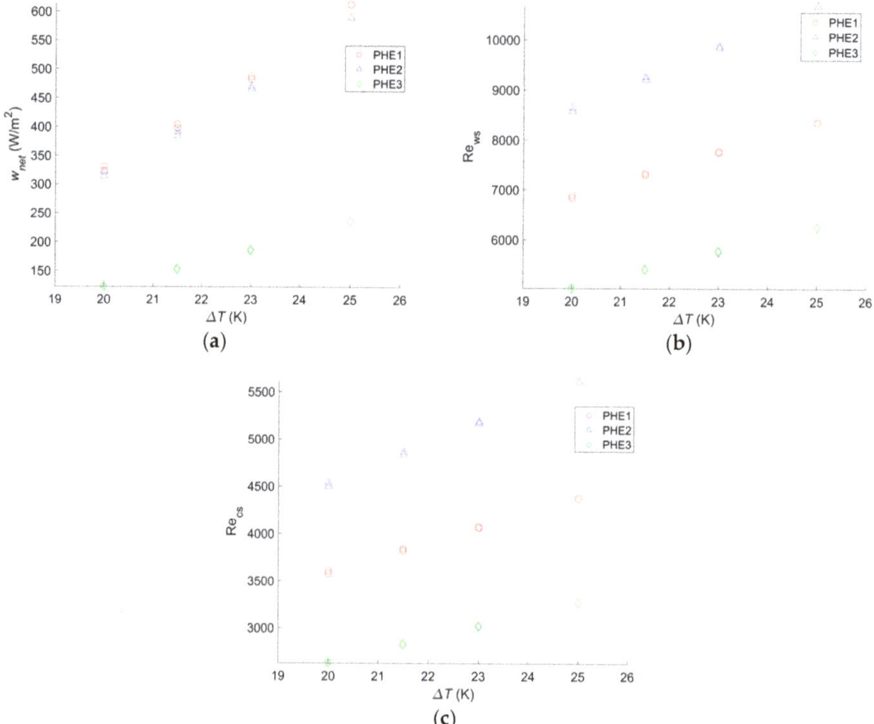

Figure 4. (**a**) Maximum net power output per square meter as a function of the temperature difference between warm and cold water, (**b**) Reynolds number of the warm water inside the evaporator as a function of the temperature difference between warm and cold water, and (**c**) Reynolds number of the cold water inside the condenser as a function of the temperature difference between warm and cold water.

$w_{net,max}$ will change significantly with a change in the seawater temperature. Indeed a 20% drop in $w_{net,max}$ was observed for the three heat exchangers when a 2 °C change in the seawater source occurred. Yeh et al. found a decrease of 35% for the same temperature drop in the cold seawater, but warm seawater was fixed at 25 °C, instead of 30 °C in the present study [30]. Sinama et al. found a 40% decrease when the warm seawater temperature dropped from 28 °C to 25 °C against 32% in this study for the same seawater temperature values [17]. Additionally, for the same temperature change, Uehara and Ikegami showed a decrease of 44% in the net power output of the OTEC power plant [23]. In their paper, VanZwieten et al. showed a decrease of 20% and 16% when the temperature changed from 20.12 °C to 21.72 °C and from 20.12 °C to 21.37 °C, respectively, against a 17% drop in this study when a 1.5 °C change occurred [31]. For a change of 5 °C in the seawater source temperature, the calculated drop in $w_{net,max}$ was 45%. The difference in the decrease of the net power output with other studies can be explained by the fact that the authors considered water ducting as well as the working fluid circulation pump [17,23,30,31]. Therefore, when the temperature difference decreases, the gross power output greatly decreases, whereas losses due to pumping and pressure drop hardly change. The impact seems to be the same if the change consists of a decrease in the warm seawater source temperature or an increase of the cold seawater source temperature.

Although the Reynolds numbers for PHE 1 were still 20% to 22% lower than those of PHE 2, the optimal operating point also varied with the temperature change; the bigger the water temperature difference is, the more an increase of the flow rate will result in an increase of the heat exchange. However, a seawater temperature change only has a limited impact on the friction factor.

Furthermore, as the maximal power output decreases with the temperature change, the range of operating points resulting in a positive net power output decreases as well.

Finally, if a change of the seawater temperature does not affect which heat exchanger leads to the highest $w_{net,max}$, the change in the operating points and the increase of the negative power output operating conditions make the flow rate monitoring even more important if the system is to be installed where the water condition changes throughout the year. In addition, for an installation where the seawater temperature does not vary, calculations need to be adjusted to the actual condition to find the optimal operating point.

4.2. Rankine Cycle Results

The results for PHE 1, PHE 2, and PHE 3, studied by Kushibe et al. [28], gave $w_{net,max}$ values that were, as expected, lower than those using the Carnot cycle, as shown in Table 4.

Table 4. Comparison between the Carnot and Rankine cycles.

	PHE 1		PHE 2		PHE 3	
	Carnot	Rankine	Carnot	Rankine	Carnot	Rankine
w_{net} (W/m^2)	613	471	590	451	237	178
Re_{ws}	8361	7876	10,690	10,121	6255	5753
V_{ws} (m/s)	0.837	0.788	1.08	1.03	0.934	0.859
Re_{cs}	4383	4381	5618	5762	3274	3277
V_{cs} (m/s)	0.832	0.831	1.08	1.11	0.927	0.928

Moreover, the results showed a difference between the optimal operating point for the same heat exchanger, especially for the Reynolds number of warm seawater. This suggests that calculations should be done for the actual cycle used in each plant to find the optimized operating point.

For this cycle as well, the calculations were realized for different temperatures, and the results are given in Figure 5.

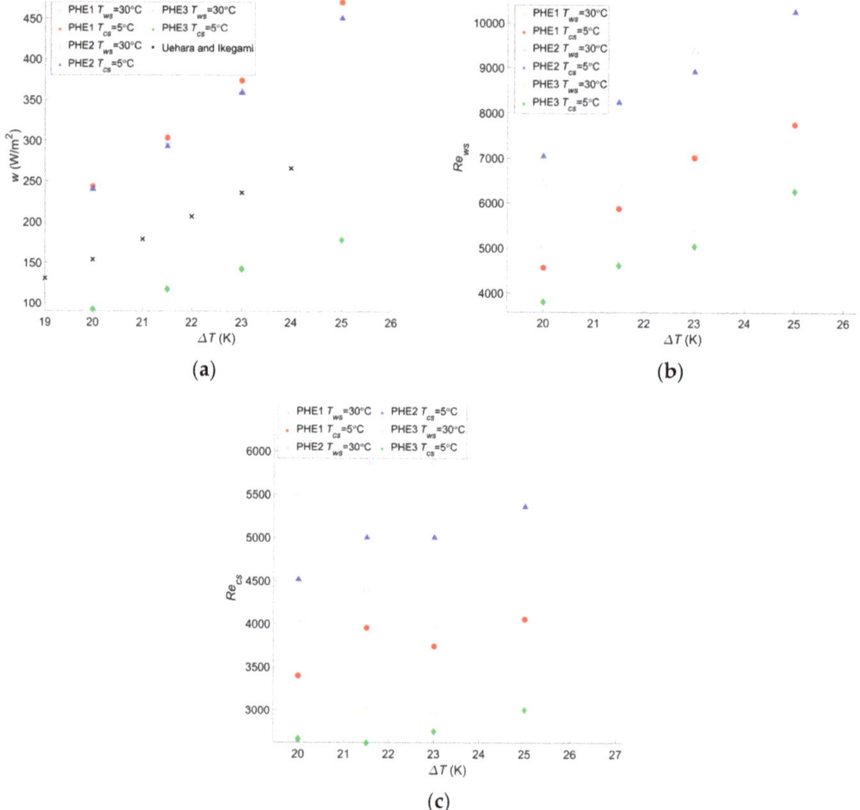

Figure 5. (**a**) Maximum net power output per square meter as a function of the temperature difference between warm and cold water, (**b**) Reynolds number of the warm water inside the evaporator as a function of the temperature difference between warm and cold water, and (**c**) Reynolds number of the cold water inside the condenser as a function of the temperature difference between warm and cold water.

These results presented a $w_{net,max}$ ranging from 90 W/m^2 to 471 W/m^2 for PHE 3 at the lowest temperature difference and PHE 1 at the highest temperature difference, respectively. As a comparison, the study performed by Uehara et al. presented a net power output of 153.9 W/m^2 at a temperature difference, ΔT, of 20 °C and a net power output of 236 W/m^2 at a ΔT of 23 °C [23]. For the same ΔT, the performances of the two heat exchangers they considered were between those of PHE 3 and PHE 2, giving a $w_{net,max}$ of 90 W/m^2 and 240 W/m^2, respectively, at a ΔT of 20 °C, and 143 W/m^2 and 360 W/m^2, respectively, at a ΔT of 23 °C. These values are rather low but are still in the range of what was found in the current study. Moreover, it should be noted that, in their paper, they considered the pumping power required for water ducting, working fluid pumps, and the heat transfer coefficient of the working fluid, which explains the lower power output.

Bernardoni et al. showed a net power output of 167 W/m^2 and a gross power output of 278 W/m^2 for a ΔT of 24 °C in their optimization using plate heat exchangers [16]. This was lower than the 293 W/m^2 found with PHE 2 at a ΔT of 21.5 °C. In their work, the authors considered water ducting, ammonia pumping, and the ammonia heat transfer coefficient, although only the latter contributed to decreasing the gross power output.

Both these studies are more accurate than the presented one; their goal was to precisely assess the net power output of the system, which was either to calculate a reliable LCOE in the case of Bernardoni et al. or for a very specific OTEC power plant design in the case of Uehara et al. The current study, however, focused on comparing the performances of heat exchangers in terms of the OTEC system performance, and such accurate assessments are less relevant in this case. In comparison, in the other two studies, there is still plenty of room for increasing the OTEC system performance regarding the choice or optimization of the heat exchanger. Regardless, the trend obtained by Uehara and Ikegami is very consistent with what was found in this work, as can be seen in Figure 5a.

As for the Carnot cycle, the results for the Rankine cycle indicate that a temperature change will have a significant impact on the maximum net power output of the OTEC power plant. A 20% decrease in $w_{net,max}$ occurred with a decrease of two degrees in the temperature difference, and a 47%–50% decrease was observed with a decrease of five degrees in the temperature difference, depending on the heat exchanger. These changes were the same as those observed with the Carnot cycle. For this cycle too, a change in the water temperature will not affect which heat exchanger is the most suitable even if the gap between each heat exchanger performance can vary along with the temperature difference. Moreover, although it is not as clear as for the Carnot cycle calculation, the Reynolds numbers of the optimized operating points tend to increase with the temperature difference. This confirms that calculations are needed for each application and its specific operating condition to find the optimal operating point. It also confirms that the flow rate should be monitored and adapted in case a change of temperature occurs. The difference between the Reynolds numbers for PHE 1 and PHE 2 remains close to the results for the Carnot cycle, with values 21% to 35% lower for PHE 1.

Although the maximum net power output and operating points vary from the Carnot cycle to another, the preferable heat exchanger remains the same. Indeed, the ratio of $w_{net,max}$ achieved using a Rankine cycle over the one achieved using the Carnot cycle was found to be fairly constant, with figures ranging between 0.73 and 0.78 for all heat exchangers and for all the computed ΔT. It is therefore possible, for the selection of the heat exchangers, to base the calculation on the Carnot cycle only.

In the case of an OTEC power plant, pressure drop plays a major role in the choice of a heat exchanger. Indeed, a study that only considers heat transfer performance might reach a different conclusion regarding which heat exchanger is the most suitable one. In such a study, the net power output would be found to increase with the Reynolds number without any limitation. Therefore, the optimum Reynolds numbers would be the maximum ones reachable for the corresponding heat exchanger and would not vary with a change in water temperature as they do with the presented results in Figures 4 and 5. The limitation induced by the pressure drop that occurs within the heat exchangers can easily be seen in the Carnot cycle results: without pressure drop consideration, the net power output cannot be negative nor decrease as the Reynolds number increases. However, Figure 3 presents parabolic graphs that include areas of negative net power output.

5. Conclusions

This work aimed to compare the performance of heat exchangers according to the OTEC system's maximum net power output. For both the Carnot and Rankine cycles, an objective function of the net power output per unit of heat exchanger surface area was defined. The objective functions took into account the pressure drop that occurred in the heat exchangers. These functions were maximized using the derivative-free method implemented in Matlab. The calculations were realized for three different heat exchangers at four temperature differences; each value of temperature difference included two points and considered an increase of the deep seawater temperature or a decrease of the warm seawater temperature. The evaluation method that was developed differs from others due to its relative ease of use. The following conclusions were made.

- For the sole purpose of a heat exchanger comparison, calculations based on the Carnot cycle for any source temperature were sufficient, as the cycle and temperature difference do not have an impact on the choice of the heat exchanger even though they do change the power output and optimized operating conditions. The Rankine cycle calculations presented a maximum net power output 23–27% lower than for the Carnot cycle that dropped ~10% each time a temperature difference decrease of 1 °C was observed. The evolution of the power output as a function of the temperature difference was found to follow the same trends as found in other studies.
- The maximum net power output was found to highly depend on the chosen heat exchangers. For the highest temperature difference, the most suitable heat exchangers among the three considered led to a maximum power output 158% and 165% higher than the worst heat exchanger for the Carnot and Rankine cycles, respectively.
- Due to the trade-off that exists between the heat transfer coefficient and the pressure drop, the heat exchanger presenting the highest heat transfer coefficient is not necessarily the one that will lead to the highest maximum net power output. In this study, for the Carnot cycle, PHE 2 competed with PHE 1 as it led to a maximum net power output that was only 3.7% lower than the one of PHE 1, even though its heat transfer coefficient was 35% lower.
- Heat exchangers with a high pressure drop and those with a low pressure drop have been found to have their own advantages and drawbacks. High pressure drop heat exchangers require lower Reynolds numbers, and therefore a smaller pumping power and/or a smaller pipe diameter are needed. Low pressure drop heat exchangers are less sensitive to a change in the Reynolds numbers, which can be useful in case a change in the operating conditions is needed. This is even more important as the results showed that negative net power output can be reached for low enough Reynolds numbers.

If the most suitable heat exchanger was found, assessments of the maximum power output and the optimum operating point are not accurate enough to be used for OTEC design. Working fluid heat transfer coefficient as well as fouling thermal resistance were neglected; however, as they can have a significant impact on the OTEC performance, they should be further studied in the future. In addition, the comparison can only be done with existing heat exchangers, as specific correlations for the heat transfer coefficient and the pressure drop are needed; thus, it is difficult to know in advance what the effect of a specific design will be. The next step is to improve the accuracy of this method by including the heat transfer coefficient of the working fluid, the pumping power for water ducting, and more specific cycles.

Author Contributions: Conceptualization, K.F. and T.Y.; methodology, K.F. and T.Y.; software, K.F.; validation, T.Y. and Y.I.; formal analysis, K.F.; investigation, K.F.; resources, T.Y. and Y.I.; data curation, K.F.; writing—original draft preparation, K.F.; writing—review and editing, K.F., T.Y., and Y.I.; visualization, K.F.; supervision, T.Y. and Y.I.; project administration, T.Y. and Y.I.; funding acquisition, T.Y. and Y.I.

Funding: This research and the APC were funded by Science and Technology Research Partnership for Sustainable Development (SATREPS), Japan Science and Technology Agency (JST)/Japan International Cooperation Agency (JICA) grant number JPMJSA1803.

Acknowledgments: This research was supported by Science and Technology Research Partnership for Sustainable Development (SATREPS), Japan Science and Technology Agency (JST)/Japan International Cooperation Agency (JICA) grant number JPMJSA1803.

Nomenclature

Nomenclature

A	[m^2]	Heat transfer surface area
As	[m^2]	Heat exchanger surface area
B	[K/W]	$\dfrac{plate\ thickness}{\lambda_{plate}A}$
Cp	[J/(kg·K)]	Specific heat
C	[W/K]	$\dot{m}Cp$ heat capacity
D	[m]	Equivalent diameter
f	[-]	Friction factor
h	[J/kg]	Specific enthalpy
L	[m]	Length of the plate
m	[kg/s]	Mass flow rate
Nu	[-]	Nusselt number
P	[kPa]	Pressure
Pr	[-]	Prandtl number
Q	[W]	Heat transfer rate
Re	[-]	Reynolds number
R_f	[m^2·K/W]	Resistance due to fouling
S	[m^2]	Total cross-sectional surface area
s	[J/(kg·K)]	Specific entropy
T	[K]	Temperature
U	[W/(m^2·K)]	Overall heat transfer coefficient
v	[m/s]	Mean velocity
Wi	[m]	Width of the plate
W	[W]	Power output
w	[W/m^2]	Power output per unit of surface area
x	[-]	Vapor quality
α	[W/(m^2·K)]	Heat transfer coefficient
ΔT	[K]	Temperature difference
λ	[W/(m·K)]	Thermal conductivity
Λ	[-]	Lagrange multiplier
μ	[Pa·s]	Dynamic viscosity
ρ	[kg/m^3]	Density
τ	[Pa]	heat exchanger wall shear stress

Subscripts

cs	Cold source
csi	Cold source inlet
cso	Cold source outlet
$gross$	Gross
in	Inlet
max	Maximum
net	Net
opt	Optimum
out	Outlet
p	Plate
w	Water
ws	Warm source
wsi	Warm source inlet
wso	Warm source outlet
wf	Working fluid

Appendix A

This appendix shows details on how the following equation can be derived,

$$T_2 = \frac{T_{wsi} - T_{wso}e^{\mathrm{NTU_{ws}}}}{\epsilon_{ws}} \quad \text{and } T_4 = \frac{T_{csi} - T_{cso}e^{\mathrm{NTU_{cs}}}}{\epsilon_{cs}} \tag{A1}$$

where T_2 and T_4 are the temperatures on the corresponding points in Figure 2.

The heat exchanged in the condenser, Qcs, is equal to:

$$Q_{cs} = mc_{p_{cs}}(T_{cso} - T_{csi}) \tag{A2}$$

Using the logarithmic mean temperature difference method, the same heat can be expressed as the following:

$$Q_{cs} = UA\Delta T_{LMTD} \tag{A3}$$

where:

$$\Delta T_{LMTD} = \frac{(T_3 - T_{cso} - (T_{csi} - T_4)}{\ln\left(\frac{T_3 - T_{cso}}{T_4 - T_{csi}}\right)} \tag{A4}$$

Given that $T_3 = T_4$, assuming the fluid is at a saturated state at the inlet and outlet of the condenser, Equation (A3) becomes:

$$Q_{cs} = UA\frac{T_{csi} - T_{cso}}{\ln\left(\frac{T_4 - T_{cso}}{T_4 - T_{csi}}\right)} \tag{A5}$$

From Equations (A2) and (A5), the equation is:

$$UA\frac{T_{csi} - T_{cso}}{\ln\left(\frac{T_4 - T_{cso}}{T_4 - T_{csi}}\right)} = -mc_{pcs}(T_{csi} - T_{cso}) \tag{A6}$$

and therefore:

$$\exp\left(\frac{UA}{mc_{pcs}}\right) = \frac{T_4 - T_{cso}}{T_4 - T_{csi}} \tag{A7}$$

$\mathrm{NTU_{cs}}$ and ϵ_{cs} are defined by [25]:

$$\mathrm{NTU_{cs}} = \frac{UA}{mC_{pcs}} \quad \text{and } \epsilon_{cs} = 1 - e^{-NTU_{cs}} \tag{A8}$$

T_4 can be deduced from Equation (A7) as:

$$T_4 = \frac{T_{csi} - T_{cso}e^{\mathrm{NTU_{cs}}}}{\epsilon_{cs}} \tag{A9}$$

For the evaporator, the assumption is that $T_1 = T_2$ is made, and then the calculations are the same. Equation (A10) is used:

$$Q_{ws} = UA\frac{(T_{wsi} - T_2 - (T_{wso} - T_1)}{\ln\left(\frac{T_{wsi} - T_2}{T_{wso} - T_1}\right)} = UA\frac{(T_{wsi} - T_{wso})}{\ln\left(\frac{T_{wsi} - T_2}{T_{wso} - T_2}\right)} = mc_{pws}(T_{wsi} - T_{wso}) \tag{A10}$$

$\mathrm{NTU_{ws}}$ and ϵ_{ws} are defined the same way:

$$\mathrm{NTU_{ws}} = \frac{UA}{mC_{pws}} \quad \text{and } \epsilon_{ws} = 1 - e^{-NTU_{ws}} \tag{A11}$$

Thus, T_2 can be expressed as:

$$T_2 = \frac{T_{wsi} - T_{wso}e^{\mathrm{NTU_{ws}}}}{\epsilon_{ws}} \tag{A12}$$

References

1. International Energy Agency. CO_2 Emissions from Fuel Combustion statistics 2017. Available online: https://www.iea.org/publications/freepublications/publication/CO2EmissionsfromFuelCombustionHighlights2017.pdf (accessed on 17 November 2019).
2. Horowitz, C.A. Paris agreement. *Int. Leg. Mater.* **2016**, *55*, 740–755. [CrossRef]
3. International Energy Agency. Key World Energy Statistics 2017. Available online: https://www.iea.org/publications/freepublications/publication/KeyWorld2017.pdf (accessed on 17 November 2019).
4. Rajagopalan, K.; Nihous, G.C. An Assessment of Global Ocean Thermal Energy Conversion Resources with a High-Resolution Ocean General Circulation Model. *J. Energy Resour. Technol.* **2013**, *135*, 041202. [CrossRef]
5. Transforming Our World: The 2030 Agenda for Sustainable Development. In *A New Era in Global Health*; Rosa, W. (Ed.) Springer Publishing Company: New York, NY, USA, 2017; ISBN 978-0-8261-9011-6.
6. Wu, C. A Performance Bound for Real OTEC Heat Engines. *Ocean Eng.* **1987**, *14*, 349–354. [CrossRef]
7. Ikegami, Y.; Bejan, A. On the Thermodynamic Optimization of Power Plants with Heat Transfer and Fluid Flow Irreversibilities. *J. Sol. Energy Eng.* **1998**, *120*, 139–144. [CrossRef]
8. Kim, A.S.; Kim, H.-J.; Lee, H.-S.; Cha, S. Dual-use open cycle ocean thermal energy conversion (OC-OTEC) using multiple condensers for adjustable power generation and seawater desalination. *Renew. Energy* **2016**, *85*, 344–358. [CrossRef]
9. Kim, A.S.; Oh, W.; Lee, H.-S.; Cha, S.; Kim, H.-J. Perspective of membrane distillation applied to ocean thermal energy conversion. *IDA J. Desalin. Water Reuse* **2015**, *7*, 17–24. [CrossRef]
10. Johnson, D. The exergy of the ocean thermal resource and analysis of second-law efficiencies of idealized ocean thermal energy conversion power cycles. *Energy* **1983**, *8*, 927–946. [CrossRef]
11. Kalina, A.I. Generation of Energy by Means of a Working Fluid, and Regeneration of a Working Fluid. U.S. Patent US4346561A, 31 August 1982.
12. Zhang, X.; He, M.; Zhang, Y. A review of research on the Kalina cycle. *Renew. Sustain. Energy Rev.* **2012**, *16*, 5309–5318. [CrossRef]
13. Uehara, H.; Ikegami, Y.; Nishida, T. OTEC System Using a New Cycle with Absorption and Extraction Processes. In *Physical Chemistry of Aqueous Systems: Meeting the Needs of Industry*; White, H.J., Jr., Sengers, J.V., Neumann, D.B., Bellows, J.C., Eds.; Begell House, Inc.: New York, NY, USA, 1995; pp. 862–869. ISBN 978-1-56700-445-8.
14. Ikegami, Y.; Yasunaga, T.; Morisaki, T. Ocean Thermal Energy Conversion Using Double-Stage Rankine Cycle. *J. Mar. Sci. Eng.* **2018**, *6*, 21. [CrossRef]
15. Dijoux, A.; Sinama, F.; Marc, O.; Clauzade, B. Working Fluid Selection General Method and Sensitivity Analysis of an Organic Rankine Cycle (ORC): Application to Ocean Thermal Energy Conversion (OTEC). 2017. hal-01653074. Available online: https://hal.archives-ouvertes.fr/hal-01653074/document (accessed on 17 November 2019).
16. Bernardoni, C.; Binotti, M.; Giostri, A. Techno-economic analysis of closed OTEC cycles for power generation. *Renew. Energy* **2019**, *132*, 1018–1033. [CrossRef]
17. Sinama, F. Thermodynamic analysis and optimization of a 10MW OTEC Rankine cycle in Reunion Island with the equivalent Gibbs system method and generic optimization program GenOpt. *Appl. Ocean Res.* **2015**, *53*, 54–66. [CrossRef]
18. Sun, F.; Ikegami, Y.; Jia, B.; Arima, H. Optimization design and exergy analysis of organic Rankine cycle in ocean thermal energy conversion. *Appl. Ocean Res.* **2012**, *35*, 38–46. [CrossRef]
19. Yasunaga, T.; Noguchi, T.; Morisaki, T.; Ikegami, Y. Basic Heat Exchanger Performance Evaluation Method on OTEC. *J. Mar. Sci. Eng.* **2018**, *6*, 32. [CrossRef]
20. Solotych, V.; Lee, D.; Kim, J.; Amalfi, R.L.; Thome, J.R. Boiling heat transfer and two-phase pressure drops within compact plate heat exchangers: Experiments and flow visualizations. *Int. J. Heat Mass Transf.* **2016**, *94*, 239–253. [CrossRef]
21. Morisaki, T.; Ikegami, Y. Performance Evaluation of Heat Exchangers in OTEC Using Ammonia/Water Mixture as Working Fluid. *Open J. Fluid Dyn.* **2013**, *3*, 302–310. [CrossRef]
22. Nilpueng, K.; Wongwises, S. Two-phase gas–liquid flow characteristics inside a plate heat exchanger. *Exp. Fluid Sci.* **2010**, *34*, 1217–1229. [CrossRef]
23. Uehara, H.; Ikegami, Y. Optimization of a Closed-Cycle OTEC System. *J. Sol. Energy Eng.* **1990**, *112*, 247–256. [CrossRef]

24. Incropera, F.P.; Lavine, A.S.; Bergman, T.L.; DeWitt, D.P. Boiling and Condensation. In *Fundamentals of Heat and Mass Transfer*; John Wiley & Sons: Hoboken, NJ, USA, 2006; pp. 619–668. ISBN 978-0-471-45728-2.

25. Incropera, F.P.; Lavine, A.S.; Bergman, T.L.; DeWitt, D.P. Heat Exchangers. In *Fundamentals of Heat and Mass Transfer*; John Wiley & Sons: New York, NY, USA, 2006; pp. 689–694. ISBN 978-0-471-47528-2.

26. Ibrahim, O.M.; Klein, S.A.; Mitchell, J.W. Effects of Irreversibility and Economics on the Performance of a Heat Engine. *J. Sol. Energy Eng.* **1992**, *114*, 267–271. [CrossRef]

27. Lemmon, E.W.; Bell, I.H.; Huber, M.L.; McLinden, M.O. *NIST Standard Reference Database 23: Reference Fluid Thermodynamic and Transport Properties-REFPROP*; Version 10.0; National Institute of Standards and Technology: Gaithersburg, MD, USA, 2018.

28. Kushibe, M.; Lkegami, Y.; Monde, M.; Uehara, H. Evaporation Heat Transfer of Amonia and Pressure Drop of Warm Water for Plate Type Evaporator. *Trans. Jpn. Soc. Refrig. Air Cond. Eng.* **2005**, *22*, 403–415.

29. Rody Oldenhuis orcid.org/0000-0002-3162-3660. "Minimize" Version 1.7, 2017/April/06. MATLAB Minimization Algorithm. Available online: http://nl.mathworks.com/matlabcentral/fileexchange/24298-minimize (accessed on 17 November 2019).

30. Yeh, R.-H.; Su, T.-Z.; Yang, M.-S. Maximum output of an OTEC power plant. *Ocean Eng.* **2005**, *32*, 685–700. [CrossRef]

31. VanZwieten, J.H.; Rauchenstein, L.T.; Lee, L. An assessment of Florida's ocean thermal energy conversion (OTEC) resource. *Renew. Sustain. Energy Rev.* **2017**, *75*, 683–691. [CrossRef]

Article

Energy and Exergy Evaluation of a Two-Stage Axial Vapour Compressor on the LNG Carrier

Igor Poljak [1,*], Josip Orović [1], Vedran Mrzljak [2] and Dean Bernečić [3]

[1] Maritime Department, University of Zadar, Mihovila Pavlinovića 1, 23000 Zadar, Croatia; jorovic@unizd.hr
[2] Faculty of Engineering, University of Rijeka, Vukovarska 58, 51000 Rijeka, Croatia; vmrzljak@riteh.hr
[3] Faculty of Maritime Studies, University of Rijeka, Studentska 2, 51000 Rijeka, Croatia; bernecic@pfri.hr
* Correspondence: ipoljak1@unizd.hr; Tel.: +385-98-613-848

Received: 8 December 2019; Accepted: 14 January 2020; Published: 17 January 2020

Abstract: Data from a two-stage axial vapor cryogenic compressor on the dual-fuel diesel–electric (DFDE) liquefied natural gas (LNG) carrier were measured and analyzed to investigate compressor energy and exergy efficiency in real exploitation conditions. The running parameters of the two-stage compressor were collected while changing the main propeller shafts rpm. As the compressor supply of vaporized gas to the main engines increases, so does the load and rpm in propulsion electric motors, and vice versa. The results show that when the main engine load varied from 46 to 56 rpm at main propulsion shafts increased mass flow rate of vaporized LNG at a two-stage compressor has an influence on compressor performance. Compressor average energy efficiency is around 50%, while the exergy efficiency of the compressor is significantly lower in all measured ranges and on average is around 34%. The change in the ambient temperature from 0 to 50 °C also influences the compressor's exergy efficiency. Higher exergy efficiency is achieved at lower ambient temperatures. As temperature increases, overall compressor exergy efficiency decreases by about 7% on average over the whole analyzed range. The proposed new concept of energy-saving and increasing the compressor efficiency based on pre-cooling of the compressor second stage is also analyzed. The temperature at the second stage was varied in the range from 0 to −50 °C, which results in power savings up to 26 kW for optimal running regimes.

Keywords: two-stage LNG compressor; energy losses; exergy destruction; energy efficiency; exergy efficiency

---

1. Introduction

The harmful pollutant emissions significantly increase over the last years in industry, marine and other energy sectors [1]. Such an increase has many negative impacts on human health as well as on the marine environment [2].

In the last years, liquefied natural gas (LNG) is recognized by many scientists and experts as a good alternative to conventional fuels due to many reasons. On one side, from the pollution point of view, the combustion of LNG removes sulfur oxides (SO_X) and particulate matter (PM) emissions almost completely, while at the same time carbon dioxide (CO_2) and nitrogen oxides (NO_X) can be significantly reduced [3,4]. On the other side, due to its low temperature (around −162 °C), LNG can be used as a heat sink in many power plants or industry processes [5–7] or its cold energy can be utilized in other applications [8]. Therefore, the usage of LNG can be beneficial in several different ways.

In the marine sector, most LNG carriers use boil-off gas (BOG) from the cargo tanks for its propulsion [9]. The domination of steam propulsion systems, which are traditionally applied on the LNG carriers [10,11] is nowadays highly influenced by dual-fuel diesel engines (DFDE) [12,13] and its possible upgrades [14]. Such dual-fuel diesel engines require a compressor unit which delivers LNG vapors from the cargo tanks to the engine.

It is very rare to find in the literature efficiency evaluation of this kind of compressor. The most scientific literature which deals with simulation and modeling of LNG systems is taking LNG (or other gas or vapor) compressor efficiency on presumptions according to the manufacturer specifications. A literature overview of several compressors is presented in Table 1, however, the type of each compressor is not known.

Table 1. Overview of compressor mass flow rate, working medium and assumed efficiencies.

Pressure	Medium	Mass Flow Rate	Simulation	Type of Compressor	Efficiency	Source
[MPa]	-	[kg/s]	-	-	η_I or η_{II} [%]	-
<1	LNG	1.39	HYSIS	Unknown	85–96	[15]
2.4	Nitrogen	1.7	Aspen	Unknown	81.98	[16]
0.55–2.5	Propane and LNG	0.9–3.44 [mol/s]	Analysis	Unknown	66–79	[17]
0.2076	Dry air	0.3134	EES	Unknown	62.7	[18]
1.305	Methane/Ethylene/Propane	3.95	HYSIS	Unknown	75	[19]
1.373	Ethane/Propane/Isobutene	2.52			85	

In some researches which deals with the LNG vapor compression, the authors clearly notify the type of compressor. As for an example, Reddy et al. [20] investigate reciprocating compressors from the LNG terminals and concluded that such compressors are notable energy consumers, which require proper optimization techniques, while Park et al. [21] analyzed dual-opposed linear compressor used in LNG (re)liquefaction process and its optimal compressor operating regimes.

In other researches, the authors investigate entire processes with LNG vapors, therefore, exact compressor types are not mentioned. As for an example, overall compression power reduction and exergy loss analysis for a cascade LNG process presented Nawaz et al. [22] where the authors found possibilities for minimization of the entire process exergy losses. Investigation of the LNG fuel gas supply system, where the compressors are used for various purposes, presented Park et al. [23]. Also in this research, the authors made several variations of the entire LNG fuel gas supply system, but the BOG compressor (or more of them) type is not presented.

The investigated DFDE LNG carrier has an electricity generating set which consists of four identical engine units. Each unit is designed to develop a total power of 7800 kW. Engines may consume diesel oil, heavy fuel oil, and LNG. LNG consuming requires pilot diesel oil burning to initiate the combustion process in LNG operation mode [24,25]. Gasified LNG is taken from the cargo tanks to the inlet of the two-stage compressor unit and delivered to the engines for completion and continuation of the combustion process [25,26]. As mentioned before, the burning of natural gas in the engines is a useful concept for reducing NO_X emissions according to Marpol Annex VI Tier 1 to Tier 3 requirements [27,28] but also it is accommodating request of maintaining cargo tank pressure in the required range, which is up to 25 kPa [29], although independent tanks class B may withstand pressures as high as 70 kPa [30]. As exploitation characteristics of this type of compressor, especially during on-board performance are rarely investigated, this paper will cover energy and exergy analysis of two-stage LNG compressor in real working conditions to better understand the power losses. Measurements were taken from the minimum flow request of the compressor, which is set from the manufacturer specifications. The engine load was gradually increased until the point of normal continuous rating at main propulsion shafts, which equals 56 rpm at each propulsion shaft.

2. Description and Characteristics of the Analyzed Two-Stage Compressor

Two-stage axial LNG compressor on LNG carrier is taking evaporated vapor from the cargo tanks, which varies in temperature at compressor inlet from about −130 to −60 °C. LNG vapor is pressurized inside the two stages to 650 kPa and delivered to the engines at a set pressure of 580 kPa [31]. Compressor capacity is controlled by variable diffuser vanes, which are adjusted according to header set pressure [31]. The temperature at the inlet of the compressor is desirable to be below −110 °C to avoid the operation of the compressor in the surge area. As previously mentioned, the suction temperature in the tank varies due to the delivery of compressor, weather conditions and pressure inside the tank. To achieve the target temperature of minimum −110 °C at the compressor inlet, the cargo spray pump injects a certain amount of liquid LNG into the suction vapor header line and in such way inlet temperature at the compressor's first stage is maintained inside a required range. The maximum electromotor rating at full speed is 610 kW and at half speed is 85 kW. These two ratings determine rotation speed which may be selected as 3550 or 1775 rpm at the electromotor side. The lower selected motor power is used only for the gradual cooling of the compressor in the recirculation loop until the targeted temperature at compressor inlet, which is below −100 °C, is reached. Once the targeted temperature is achieved compressor has permission to start at a higher speed. Half speed is not in use in normal navigation mode, due to the low delivery pressure of the compressor to the engines and this option is useless for this type of engine. However, half a rotation speed option may be used for burning excess boil-off gas from cargo tanks in the gas combustion unit. The compressor is driven by an electromotor which is coupled to spur reduction gear. Particulars of the two-stage compressor are given in Table 2 [32].

Table 2. Two stage compressor main design particulars.

Two Stage Compressor Main Characteristics	Operation Parameter
Capacity at full speed	3200 m^3/h
Capacity at half speed	520 m^3/h
Suction pressure	104 kPa
Suction temperature	−120 °C
Discharge pressure at full speed	700 kPa
Discharge pressure at half speed	160 kPa
Shaft rpm at full speed	29,775 rpm
Shaft rpm at half speed	14,888 rpm
Electric motor power	610 kW

The schematic outline of the two-stage compressor is given in Figure 1, where p_1, t_1 is LNG vapor pressure and temperature at compressor inlet to the first stage. Delivery pressure and temperature from the compressor first stage and the inlet to the compressor second stage are p_2 and t_2, respectively. Outlet pressure and temperature from the second stage are p_3 and t_3, respectively. The mass flow rate ($\dot{m}$) is the same through all operation measurement range.

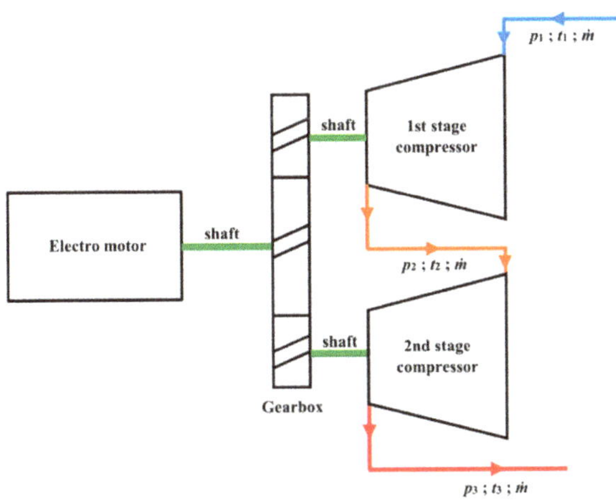

Figure 1. Two-stage compressor outline with given flows.

The energy efficiency data for given conditions, provided by the manufacturer are presented in Table 3.

Table 3. Two-stage compressor design operating parameters [33].

Parameter	Unit	1	2	3	4
Gas		CH$_4$	90CH$_4$/10N$_2$	BOG	Half speed
Inlet pressure	[kPa]	104	104	104	104
Outlet pressure	[kPa]	700	700	700	700
Inlet temperature	[°C]	−120	−120	−120	−120
Outlet temperature	[°C]	41.5	55.4	42.7	−75.9
Energy efficiency	[%]	55	52.1	55.2	39
Coupling power	[kW]	485	514	488	68

3. Measuring Equipment and Measuring Results

Sea trial tests were carried out at different main propulsion shaft rotation speed. The test commenced at the lowest load of the two-stage compressor, where the compressor is delivering all gas to the generator engines, which is at about 46 rpm at the main propulsion shafts. Measuring results (Tables 4 and 5) were obtained with ship's standard measuring equipment [34–36] for various readings which include pressure, temperature and mass flow rate of LNG vapor at compressor first stage inlet, first stage outlet and second stage outlet. Before commencing the sea trial test, it was required to run diesel generators on the low sulfur diesel oil which is not the preferred option for the charterer due to fuel cost. Once when cooling down temperature is achieved, two-stage compressor received permission to start at higher motor speed. Change over process from diesel oil combustion to gas combustion was carried out according to the producer pre-set sequence. Fuel oil consumption measuring results were carried out during the parallel operation of three and four generators online.

Table 4. Two-stage compressor measured pressure and temperatures with main propulsion shafts speed variations.

Port Side Shaft Output	Starboard Side Shaft Output	First Stage Inlet Pressure	First Stage Inlet Temperature	First Stage Outlet Pressure	First Stage Outlet Temperature	Second Stage Outlet Pressure	Second Stage Outlet Temperature
[rpm]	[rpm]	[kPa]	[°C]	[kPa]	[°C]	[kPa]	[°C]
46.05	46.58	112.5	−109.2	355	−13	684.7	75.2
47.16	47.18	112.2	−113.9	357.8	−15	695.5	75.8
48.17	48.2	111.8	−115.9	340	−16.8	665.9	74.5
49.56	49	111.5	−114.3	341.4	−16.2	674.9	74
50.88	49.81	109.9	−113.8	341.3	−16.6	656.5	73.3
51.21	50.82	109.7	−110.8	345	−15.7	688.1	72.3
52.83	51.64	109.9	−119	347.8	−20.6	682.6	69.6
53.61	52.21	109.1	−116.5	335.7	−20.1	661.2	68.7
54.54	53.42	108.3	−114	336.6	−19.6	671.6	67.6
55.63	54.76	108	−114	335	−20.1	675	66.8
55.37	56.33	106.8	−109.6	321.5	−18.2	658.7	67

Table 5. Two-stage compressor measured mass flow rate, voltage, amperage and power factor with main propulsion shafts speed variations.

Port Side Shaft Output	Starboard Side Shaft Output	Mass Flow Rate	Voltage	Amperage	Measured Power Factor
[rpm]	[rpm]	[kg/h]	[V]	[A]	[φ]
46.05	46.58	2320	6764	40.4	0.85
47.16	47.18	2535	6763	41.7	0.85
48.17	48.2	2650	6759	41.3	0.85
49.56	49	2741	6756	41.9	0.85
50.88	49.81	3150	6764	45.4	0.85
51.21	50.82	3291	6763	46	0.85
52.83	51.64	3382	6759	45.2	0.85
53.61	52.21	3509	6758	47.4	0.85
54.54	53.42	3672	6752	49.9	0.85
55.63	54.76	3792	6755	50.3	0.85
55.37	56.33	3926	6752	51.3	0.85

Equipment for electrical readings of voltage, ampere, and power factor is described in [37], and an overview of propulsion shaft revolutions measurements can be found in the literature [38,39]. Measured results are given in Table 5. The power factor of the compressor electromotor varied only for negligible values throughout all measured range at the high voltage control and measuring breaker. Voltage is not changeable in the power control process and is coming from the common ship's power network, which is 6600 V for this type of vessel. Power network is supplied from four diesel generator units which have the same power output of 7800 kW. For the maneuvering purposes, it is required to have two units to run in parallel, although the power consumption may withstand one unit until dead slow ahead and dead slow astern operation is required. This option is preferred due to safety reasons. Although set voltage should be 6600 V it may be seen from Table 5 that shipyard increased set point for about 2% during the sea trials. The high outlet pressure from the second stage compressor outlet is required due to pressure drops in the gas line to the engines in order to satisfy the engine gas pressure request of 580 kPa. Increasing power at the compressor's electromotor is achieved by increasing amperage with variable frequency drive [40–42]. Measured results with compressor load variations are given in Table 5.

Loaded LNG content (molar ratio) is [43]: CH_4 (0.9494), C_2H_6 (0.0475), C_3H_8 (0.0017), i-C_4H_{10} (0.0001), n-C_4H_{10} (0.0001), N_2 (0.0012).

4. Thermodynamic Efficiency Analysis

Two-stage compressor is evaluated for energy efficiency according to recommendations in the following literature [44–48]. Energy efficiency formulation with referring to Figure 1 is given below.

The mass flow rate balance for a general steady-flow system is given as:

$$\sum_{IN} \dot{m} = \sum_{OUT} \dot{m} \tag{1}$$

Energy rate balance for a general steady-flow system is:

$$\dot{Q}_{IN} + P_{IN} + \sum_{IN} \dot{m} \cdot \left(h + \frac{\tilde{v}^2}{2} + g \cdot z \right) = \dot{Q}_{OUT} + P_{OUT} + \sum_{OUT} \dot{m} \cdot \left(h + \frac{\tilde{v}^2}{2} + g \cdot z \right) \tag{2}$$

In the steady flow processes, the total mass and energy content remains constant in the control volume, thus the total energy rate change of the system is zero:

$$\dot{E}_{IN} - \dot{E}_{OUT} = \frac{d\dot{E}}{dt} = 0,$$

(3)

and then

$$\dot{E}_{IN} = \dot{E}_{OUT}.$$

(4)

Energy efficiency is a ratio of useful and used energy streams in the process:

$$\eta_I = \frac{\dot{E}_{OUT}}{\dot{E}_{IN}} = 1 - \frac{\dot{E}l}{\dot{E}_{IN}}.$$

(5)

The irreversible adiabatic process is calculated as per equations from (6) to (8) in order to match with the producer's results given in Table 3. A compression process with ideal or adiabatic compression and the actual (polytropic) process being an irreversible adiabatic process is shown in the h–s diagram in Figure 2.

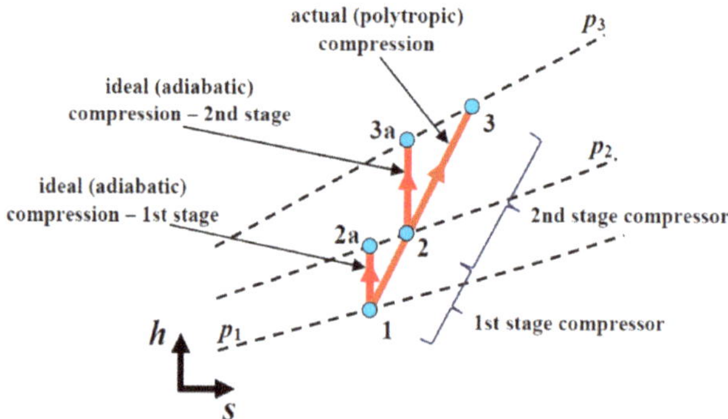

Figure 2. The *h–s* diagram of the compression process in the two-stage compressor.

Energy efficiency of the compressor first stage is:

$$\eta_{I_1} = \frac{P_{1adiabatic}}{P_{1actual}} = \frac{\dot{m}\cdot(h_{2a} - h_1)}{\dot{m}\cdot(h_2 - h_1)} = \frac{(h_{2a} - h_1)}{(h_2 - h_1)}.$$

(6)

Energy efficiency of the compressor second stage is:

$$\eta_{I_2} = \frac{P_{2adiabatic}}{P_{2actual}} = \frac{\dot{m}\cdot(h_{3a} - h_2)}{\dot{m}\cdot(h_3 - h_2)} = \frac{(h_{3a} - h_2)}{(h_3 - h_2)}.$$

(7)

Overall two-stage compressor energy efficiency (according to Figure 2) is:

$$\eta_I = \frac{P_{adiabatic}}{P_{actual}} = \frac{\dot{m}\cdot(h_{2a} - h_1) + \dot{m}\cdot(h_{3a} - h_2)}{\dot{m}\cdot(h_3 - h_1)} = \frac{(h_{2a} - h_1) + (h_{3a} - h_2)}{(h_3 - h_1)}.$$

(8)

Actual measured temperatures, pressures, and mass flow rate are used for calculating the actual (polytropic) power for running the compressor's first stage only:

$$P_{1actual} = \dot{m}\cdot(h_2 - h_1),$$

(9)

and for running the compressor's second stage only:

$$P_{2actual} = \dot{m} \cdot (h_3 - h_2). \tag{10}$$

Mass flow rate at the compressor first stage is:

$$\dot{m} = \frac{P_{1actual}}{(h_2 - h_1)}. \tag{11}$$

Mass flow rate at the compressor second stage is:

$$\dot{m} = \frac{P_{2actual}}{(h_3 - h_2)}. \tag{12}$$

Equalling Equations (11) and (12) gives:

$$P_{2actual} = P_{1actual} \cdot \frac{(h_3 - h_2)}{(h_2 - h_1)}. \tag{13}$$

The actual power required for running the whole compressor is:

$$P_{actual} = P_{1actual} + P_{2actual} = P_{1actual} + P_{1actual} \cdot \frac{(h_3 - h_2)}{(h_2 - h_1)} = P_{1actual} \cdot \left[\frac{(h_3 - h_2)}{(h_2 - h_1)} + 1 \right] = P_{1actual} \cdot \frac{(h_3 - h_1)}{(h_2 - h_1)}. \tag{14}$$

For the energy losses, exergy efficiency and exergy destruction calculation the idea is to present all power losses from the electrical power input, which is required to run the compressor electromotor until it's output. For such analysis, it was required to split the measured electrical power at the common electromotor inlet to the first and second stages of the compressor. The splitting of the power done through the proportion of power at the actual compression irreversible process. Formula for calculation of electric motor power from measured results is drawn from literature [49,50]:

$$P_{EM} = \sqrt{3} \cdot I \cdot U \cdot \cos \varphi. \tag{15}$$

Total electric power for the running of the first and the second compressor stage, which includes all irreversibility's inside the process such are: electrical losses of the electric motor, mechanical losses inside the power transmission reduction gears, power losses in the shaft seals, etc. equals to:

$$P_{EM} = P_{1EM} + P_{2EM}. \tag{16}$$

Electrical energy distributed to power each compressor stage (first and second) is divided in the same proportions as actual (polytropic) power. Consumed electrical energy for the running of the compressor first stage is then:

$$P_{1EM} = P_{EM} \cdot \frac{(h_2 - h_1)}{(h_3 - h_1)}. \tag{17}$$

Consumed electrical energy for the running of the compressor second stage is:

$$P_{2EM} = P_{EM} - P_{1EM}. \tag{18}$$

Energy rate balance of the compressor first stage is:

$$\dot{m} \cdot h_1 + P_{1EM} = \dot{m} \cdot h_2 + \dot{El}_1. \tag{19}$$

Energy rate losses of the compressor first stage are:

$$\dot{El}_1 = \dot{m} \cdot (h_1 - h_2) + P_{1EM}. \tag{20}$$

Energy rate balance of the compressor second stage is:

$$\dot{m}{\cdot}h_2 + P_{2EM} = \dot{m}{\cdot}h_3 + \dot{El}_2. \tag{21}$$

Energy rate losses of the compressor second stage are:

$$\dot{El}_2 = \dot{m}{\cdot}(h_2 - h_3) + P_{2EM}. \tag{22}$$

Energy rate balance of the two-stage compressor is:

$$\dot{m}{\cdot}h_1 + P_{1EM} + \dot{m}{\cdot}h_2 + P_{2EM} = \dot{m}{\cdot}h_2 + \dot{m}{\cdot}h_3 + \dot{El}. \tag{23}$$

Overall energy rate losses of the two-stage compressor are:

$$\dot{El} = \dot{m}{\cdot}(h_1 - h_3) + P_{EM}. \tag{24}$$

Exergy evaluation of two-stage compressor is performed according to the literature [51–59]: exergy rate balance for steady flow systems is:

$$\sum_{IN}\left(\dot{m}{\cdot}ex\right) + \sum\left(1 - \frac{T_0}{T}\right){\cdot}\dot{Q} = \sum_{OUT}\left(\dot{m}{\cdot}ex\right) + P + T_0{\cdot}\Delta S, \tag{25}$$

where specific exergy is given by:

$$ex = h - h_0 - T_0{\cdot}(s - s_0). \tag{26}$$

Exergy rate balance of the compressor first stage is:

$$\dot{m}{\cdot}ex_1 + P_{1EM} = \dot{m}{\cdot}ex_2 + \dot{Exd}_1. \tag{27}$$

Exergy destruction rate of the compressor first stage is:

$$\dot{Exd}_1 = \dot{m}{\cdot}(ex_1 - ex_2) + P_{1EM}. \tag{28}$$

Exergy efficiency of the compressor first stage is:

$$\eta_{II_1} = 1 - \frac{\dot{Exd}_1}{\dot{Ex}_{IN}} = \frac{\dot{Ex}_{OUT}}{\dot{Ex}_{IN}} = 1 - \frac{\dot{m}{\cdot}ex_1 - \dot{m}{\cdot}ex_2 + P_{1EM}}{P_{1EM}} = \frac{\dot{m}{\cdot}ex_2 - \dot{m}{\cdot}ex_1}{P_{1EM}}. \tag{29}$$

Exergy rate balance of the compressor second stage is:

$$\dot{m}{\cdot}ex_2 + P_{2EM} = \dot{m}{\cdot}ex_3 + \dot{Exd}_2. \tag{30}$$

Exergy destruction rate of the compressor second stage is:

$$\dot{Exd}_2 = \dot{m}{\cdot}(ex_2 - ex_3) + P_{2EM}. \tag{31}$$

Exergy efficiency of the compressor second stage is:

$$\eta_{II_2} = 1 - \frac{\dot{Exd}_2}{\dot{Ex}_{IN}} = \frac{\dot{Ex}_{OUT}}{\dot{Ex}_{IN}} = 1 - \frac{\dot{m}{\cdot}ex_2 - \dot{m}{\cdot}ex_3 + P_{2EM}}{P_{2EM}} = \frac{\dot{m}{\cdot}ex_3 - \dot{m}{\cdot}ex_2}{P_{2EM}}. \tag{32}$$

Exergy rate balance of two-stage compressor is:

$$\dot{m}{\cdot}ex_1 + P_{1EM} + \dot{m}{\cdot}ex_2 + P_{2EM} = \dot{m}{\cdot}ex_2 + \dot{m}{\cdot}ex_3 + \dot{Exd}. \tag{33}$$

Exergy destruction rate of the two-stage compressor is:

$$\dot{Ex}d = \dot{m}\cdot(ex_1 - ex_3) + P_{EM}. \tag{34}$$

Exergy efficiency of the two-stage compressor is:

$$\eta_{II} = 1 - \frac{\dot{Ex}d}{\dot{Ex}_{IN}} = \frac{\dot{Ex}_{OUT}}{\dot{Ex}_{IN}} = 1 - \frac{\dot{m}\cdot ex_1 - \dot{m}\cdot ex_3 + P_{EM}}{P_{EM}} = \frac{\dot{m}\cdot ex_3 - \dot{m}\cdot ex_1}{P_{EM}}. \tag{35}$$

The assumed atmospheric condition for exergy calculation is 25 °C and 0.1 MPa according to the literature overview [60,61].

5. Energy and Exergy Analysis Results and Discussion

Calculated electromotor power distribution between two compressors are presented in Table 6. As previously mentioned, the LNG carrier has two propulsion shafts and for convenience only average measured rpm of both shafts will be given through the discussion. As can be seen from Table 6, the two-stage compressor is well balanced and load shearing between the two stages is within 0.5%. The first stage is slightly more loaded compared to the second stage. The increase of compressor electromotor power with load variation is about 107 kW. That is a relatively small amount, compared to total power required for a two-stage compressor, which is required for running the unit at the lowest delivery zones.

Table 6. Two-stage compressor measured electromotor power and calculated load shearing ratio between two compressors.

Main Engines Shaft Output	First Stage Calculated Power	Second Stage Calculated Power	Measured Electromotor Power	First Stage Load Share	Second Stage Load Share
[rpm]	[kW]	[kW]	[kW]	[%]	[%]
46.32	201.99	200.33	402.31	50.21	49.79
47.17	208.27	206.93	415.20	50.16	49.84
48.19	206.12	204.85	410.97	50.15	49.85
49.28	209.17	207.59	416.76	50.19	49.81
50.35	226.18	225.92	425.10	50.03	49.97
51.02	229.24	228.77	458.01	50.05	49.95
52.24	226.76	223.02	449.78	50.42	49.58
52.91	237.06	234.54	471.60	50.27	49.73
53.98	248.94	247.10	496.04	50.19	49.81
55.20	250.92	249.32	500.23	50.16	49.84
55.85	254.98	254.97	509.95	50.00	50.00

Two-stage compressor energy loss is shown in Figure 3. As the rotation speed of the main propulsion shaft increases, more gas has to be delivered to the generator engines. A higher load at generators is proportional to the withdrawal of more vapor from the cargo tanks. The side effect is that cargo tank pressure is reduced and colder cargo vapor is flowing to the compressor inlet. That process continues and the temperature is going down at rotation speed of main propulsion shafts from 46.3 to 52.2 rpm. Increased mass flow rate and decreased temperature at compressor inlet reduce energy losses and a peak of energy loss reduction is at 52.2 rpm of main propulsion shafts. After 52.2 rpm at main propulsion shafts temperature at compressor's first stage outlet stabilizes and losses become steady at about 50 kW at each compressor stage, Figure 3. Total energy loss of compressor is higher in the lower running zones and varies from 80 to 150 kW for both compressor stages. Energy losses of the first and the second compressor stages are practically the same through all measured range what is shown in Figure 3.

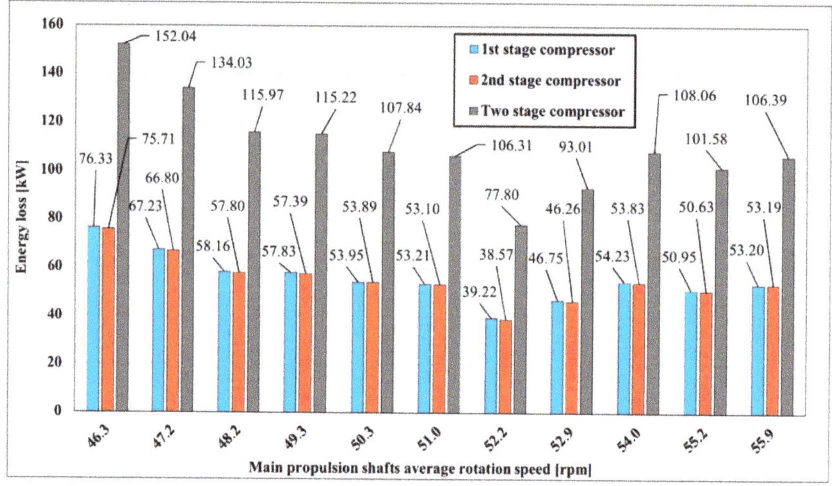

Figure 3. Two-stage compressor energy loss with main propulsion shafts speed variation.

The energy efficiency of the compressor is shown in Figure 4. The compressor energy efficiency of the first stage is higher compared to the second stage and is in the average range of about 52%. The energy efficiency of the second stage average range is up to 48%. As compressor increases delivery, energy efficiencies of both stages converge and are for both stages above 50%. This effect is connected to reduced offset from optimal intermediate pressure at the outlet of the first compressor stage [57]. The optimal intermediate pressure is calculated as [57]:

$$p_2 = \sqrt{p_1 \cdot p_3},$$

(36)

where:

p_1—is pressure at the compressor inlet to the first stage [kPa],
p_2—is pressure at the compressor outlet from the first stage [kPa],
p_3—is pressure at the compressor outlet from the second stage [kPa].

The offset is:

$$\Delta p = p_2 - \sqrt{p_1 \cdot p_3}.$$

(37)

As the two-stage compressor mass flow rate becomes higher offset decreases what acts beneficially to the two-stage compressor energy efficiency.

The exergy destruction of the two-stage compressor is higher comparing to energy losses where peak destruction is raised to about 320 kW. It may be seen in Figure 5 that the first stage compressor destructs more exergy compared to the second compressor stage in all measured ranges, which was not clearly visible in energy analysis. As the exergy destruction of the first stage compressor is higher compared to the second stage compressor, the efficiency of the first stage compressor will be lower. Destructed exergy of the first stage compressor is from 150 kW at a rotation speed of main propulsion shaft 46.3 rpm and is gradually increased to about 170 kW at 55.9 rpm, whereas destructed exergy of the second stage compressor is almost equally distributed around 140 kW, as shown in Figure 5.

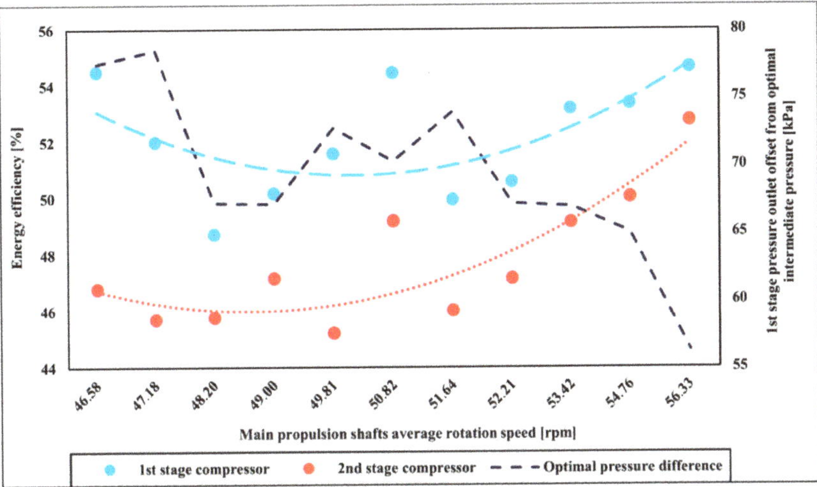

Figure 4. Two-stage compressor energy efficiency with main propulsion shafts speed variation.

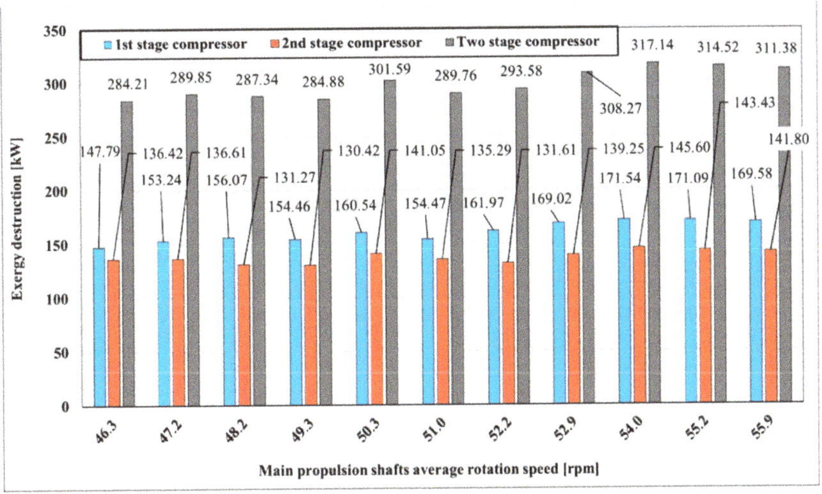

Figure 5. Two-stage compressor exergy destruction with main propulsion shafts speed variation.

The outcome of exergy destruction analysis is the higher exergy efficiency of the second stage compressor comparing to the first stage compressor in all measured range. The first stage compressor exergy efficiency varies with the compressor load from about 24 to 34% at the measured range. The best running conditions are achieved at a rotation speed of 51 and 55.9 rpm at the main shafts. The second stage of the compressor has a higher exergy efficiency, which varies from about 32 to 45%, but contrary to the first stage of the compressor best results are distributed at a rotation speed of 51 rpm onwards at the main propulsion shafts. By increasing the delivery of the compressor, exergy efficiency increases, however at the same time efficiency offset between two compressors stages becomes higher as well. Optimal intermediate pressure does not have the same effect on exergy efficiency as on energy efficiency. In fact, it is the opposite. As previously explained, also shown in Figure 6, a discrepancy in the exergy efficiency of the first and second stages is about 5% at the beginning, and at the end of the measured range, the discrepancy increases to about 10%. It means that decreasing the optimal intermediate pressure offset exergy efficiency of both compressor stages has better results, but discrepancy in efficiency increases as well. Exergy efficiency is increased smoothly in entire ranges for the second stage of the compressor. Regarding the first stage of the compressor, exergy efficiency is increasing in the whole measured range, but it is not evenly distributed as it is distributed in the second stage.

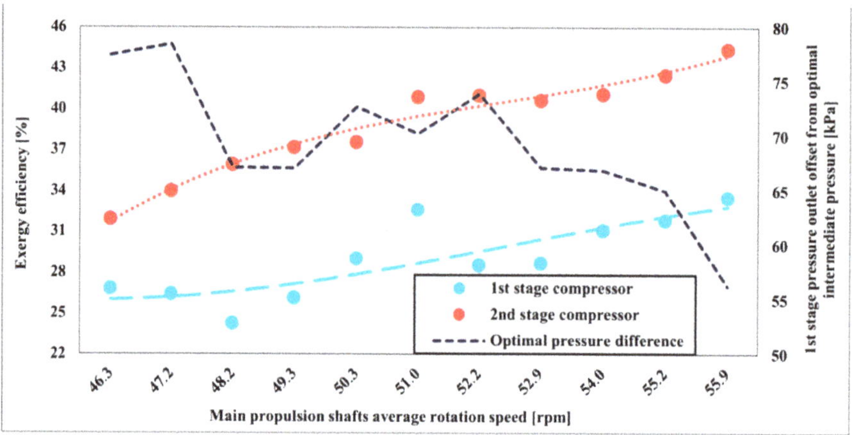

Figure 6. Two-stage compressor exergy efficiency with main propulsion shafts speed variation.

The overall efficiency of the two-stage compressor is given in Figure 7. The energy efficiency of the two-stage compressor in the observed range, in the beginning, is 50.7%. As the rotation speed of the main propulsion shaft increases, energy efficiency is dropping to about 47.3% at 48.2 rpm. The average measured energy efficiency in the whole range is 49.9% and the best energy efficiency is 53.7% at the last measured propulsion shaft rotation speed. Measured results are, on average, about 5% lower compared to the maker's efficiency. Although calculated efficiency is lower compared to the producer results it has to be noticed that the compressor was running on the lower delivery pressure and higher inlet temperature at the first compressor stage what is affecting compressor efficiency. It could be concluded that calculated efficiency will be even closer to the producer results as running parameters converge to the parameters given by the producer. On the other side, exergy efficiency is lower, about 29% in the beginning and with the increased compressor load overall exergy efficiency increases to about 39%. The average exergy efficiency of the two-stage compressor is 33.9%. As exergy efficiency is taking entropy generation into account it is clearly seen that two-stage cryogenic compressor is not as efficient as it appears through the energy efficiency analysis.

The exergy efficiency of the LNG compressor is influenced by the variation of atmospheric temperature. The change of exergy efficiency in dependence on the outside temperature range from 0 to 50 °C is analyzed. The mentioned temperature range is chosen because these two temperatures are extreme temperatures that the vessel will operate most of the time. The referent temperature for the exergy calculation is 25 °C and 0.1 MPa. Figure 8 shows the first stage compressor's exergy efficiency changes with temperature variation. As can be seen in Figure 8, higher exergy efficiency is achieved at lower reference temperatures. As temperature increases, exergy efficiency decreases by about 7.7% on average over the whole measured range. The lowest sensitivity to the reference temperature change is in the lower running main propulsion shaft rotation speed where efficiency decreases around 1.18% for every 10 °C from 0 to 50 °C. The highest temperature sensitivity is at 52.83 rpm at the main propulsion shaft rotation speed where it is changing at the rate of 1.81%.

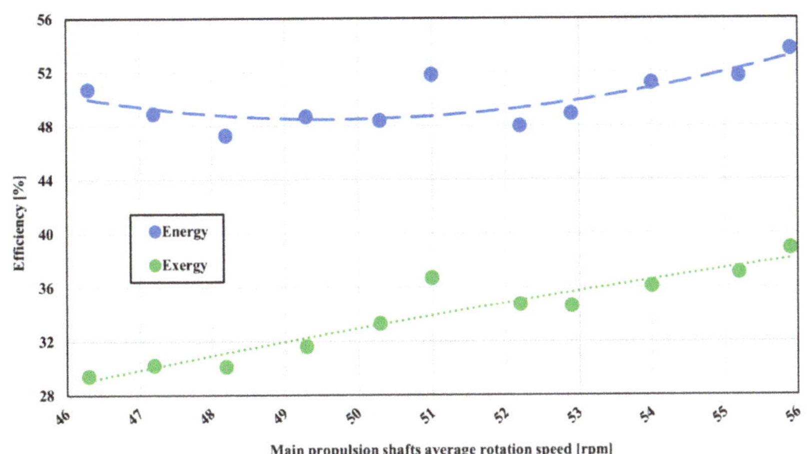

Figure 7. Two-stage compressor energy and exergy efficiency with main propulsion shafts speed variation.

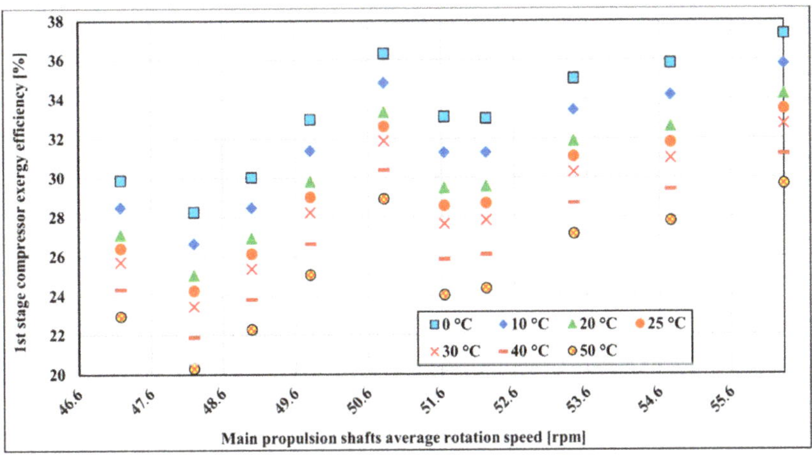

Figure 8. First stage compressor exergy efficiency with surrounding temperature change.

The second compressor stage is less influenced by the surrounding temperature variation compared to the first compressor stage. Exergy changes about 6.1% on average over the measured range (Figure 9). The temperature sensitivity distribution is similar to the first compressor stage and is the lowest in the lower running main propulsion shaft rotation speed where exergy efficiency decreases around 1.01% for every 10 °C from 0–50 °C. The highest temperature sensitivity is at 52.83 rpm of main propulsion shaft rotation speed where it is changing at the rate of 1.39%.

Overall compressor exergy efficiency changes are 6.9% on average for the whole measured range with surrounding temperature variations from 0 to 50 °C (Figure 10). The temperature sensitivity distribution is proportional to the compressor first and second stage exergy efficiency change. According to overall results, it is preferable for a compressor to operate at the lowest atmospheric temperatures in order to have better exergy efficiency. However, LNG carriers are sailing all over the world in different climates and ship's crew and/or compressor operators cannot influence at atmospheric conditions in order to increase the exergy efficiency but this analysis can help in better understanding of compressor processes.

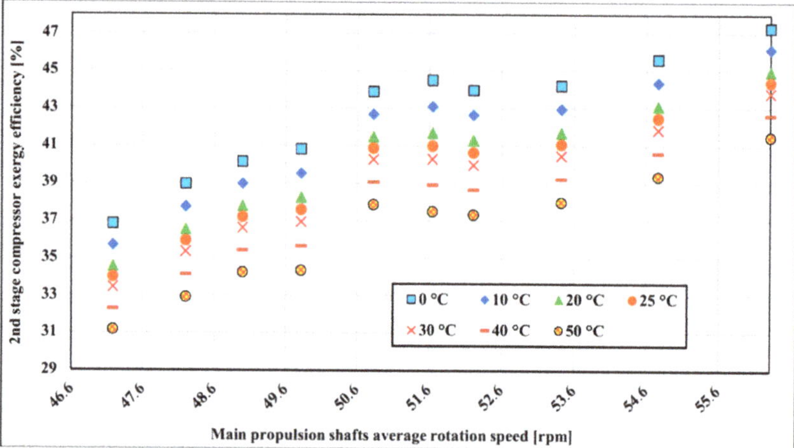

Figure 9. Compressor's second stage exergy efficiency with surrounding temperature change.

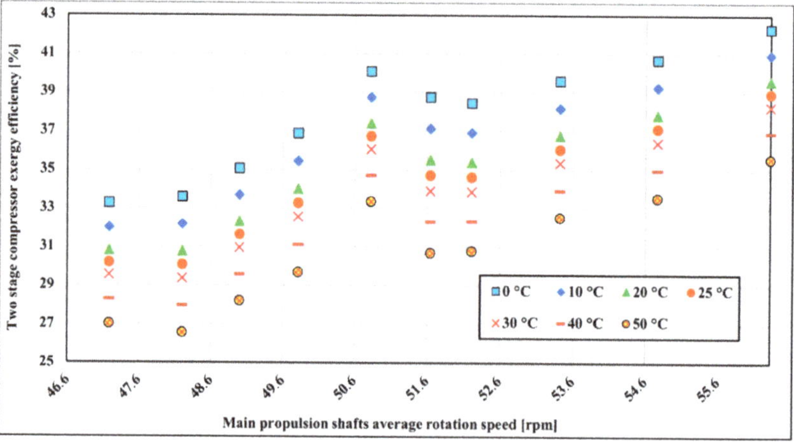

Figure 10. Two-stage compressor exergy efficiency with surrounding temperature change.

6. Variation and Optimization of Temperature at the Compressor's Second Stage Inlet

In this part, the variation and optimization of temperature at the compressor's second stage are presented. The proposed new concept of pre-cooling the vapor before the compressor second stage could enable energy saving and increase compressor efficiency.

Part of LNG may be taken from the cargo tank with the purpose to cool pressurized LNG before the compressor's second stage. Such a scenario will not require forcing vaporizer in the system. When the ship is sailing on the calm sea and boil-off gas is not sufficient in the cargo vapor header part of this vaporized liquid after-cooler will evaporate and maintain sufficient pressure in the cargo tanks. The purpose of forcing vaporizer is to vaporize part of the liquid from the cargo tanks and to maintain compressor suction pressure in the allowed ranges and to avoid reducing cargo tank pressure excessively. If cooler is installed between two stages of compressor this could partially replace the forcing vaporizer. Cooling between two stages may be achieved either with additional cooler or with direct injection of part of the liquid inside the outlet of the first stage compressor by the spray pump.

Actual (polytropic) power saving depends on the compressor's second stage inlet temperature is given in Figure 11. The temperature range from 0 to −50 °C is chosen. The energy efficiency of the two-stage compressor had been taken from the previous calculation for the various purposes and is set as the fixed value. The cumulative calculated power needed for increasing pressure to measured set pressure of 650 kPa at different loads is compared to actual measured power by changing inlet temperatures of the second stage. Running off the main propulsion shaft from 46.5 to 56.3 rpm is giving a cumulative power saving of two-stage compressor from 7 to 25.8 kW in the temperature range from −30 to −50 °C at compressor inlet to the second stage. For the pre-cooling in the range from −20 to −30 °C at compressor inlet to the second stage cumulative power savings are in the range from 0.3 to 10 kW but from 51.64 to 54.76 rpm at −20 °C at compressor inlet to the second stage there is no cumulative power savings. Temperatures above −20 °C at compressor inlet to the second stage give negative results and the compressor is consuming more power in such a working regime in all running zones.

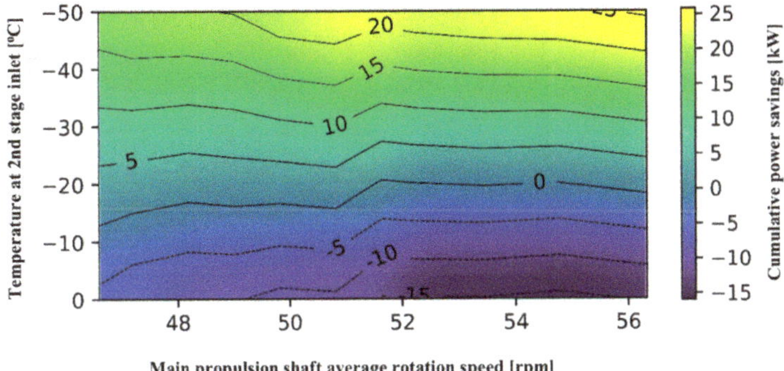

Figure 11. Two-stage compressor cumulative power optimum saving with second stage inlet temperature variation.

7. Conclusions

In this paper, the energy and exergy evaluation of the two-stage axial vapor compressor is presented. Data from cryogenic compressors on the dual-fuel diesel–electric LNG carrier were measured and analyzed to investigate compressor energy and exergy efficiency in real exploitation conditions. Operating parameters of this type of compressor, especially during on-board performance are rarely investigated, so the presented analysis enables insight into better understanding the compressor processes in different operating regimes. The results show how main engine load variation has an influence on compressor performance which is not found in the producers' manuals or scientific literature.

Energy efficiency of compressor first and second stages has the highest value at the lowest main engine rotation speed and at the highest delivery ranges. The presented results show that the overall average calculated energy efficiency of the two-stage compressor is 49.9% and the best energy efficiency is about 54% at the last measured propulsion shaft rotation speed. At the average rotation speed of the main propulsion shaft of 48.2 rpm, energy efficiency decreases to the lowest value of about 47%. Calculated efficiency is lower compared to the producer results but it has to be noticed that the compressor was running on the lower delivery pressure and higher inlet temperature at the first compressor stage what is affecting compressor efficiency. Producers usually present only the best results and at full compressor capacity but in this paper, the whole range of compressor regime was analyzed. Although the compressor first and second stages are well balanced, the first stage will achieve higher efficiency if the pressure rise offset is lower.

Compressor exergy efficiency is lower, about 29% in the beginning and with the increasing compressor load overall exergy efficiency increases to about 39%. The average exergy efficiency of the two-stage compressor is 33.9%. However, compressor manufacturers never evaluate the exergy efficiency of the compressor as its efficiency is lower compared to energy efficiency. According to exergy analysis, the two-stage compressor is not very efficient equipment and there is still room for design improvements.

Examination of the exergy efficiency in dependence on surrounding temperature shows that the two-stage compressor is more effective at lower temperatures. In that respect, the two-stage compressor is behaving similarly to the steam turbines [61,62]. The compressor first stage is more sensitive to temperature changes as it works with higher temperature differences than the second stage. Overall compressor exergy efficiency changes are 6.9% on average for the whole measured range with surrounding temperature variations from 0 to 50 °C. Temperature sensitivity analysis shows that the lowest compressor first stage exergy efficiency decrease is around 1.18% with every 10 °C from 0 to 50 °C at the lowest measured main propulsion shaft average rotation speed. The highest temperature sensitivity is at 52.83 rpm at the main propulsion shaft rotation speed where it is changing at the rate of 1.81%. The lowest second compressor stage exergy efficiency change is at lowest measured main propulsion average speed around 1.01% with every 10 °C from 0 to 50 °C and the highest temperature sensitivity is at 52.83 rpm of main propulsion shaft rotation speed where it is changing at the rate of 1.39%.

The proposed new concept of inter-cooling of the LNG after the first stage of the compressor could increase the efficiency of the compressor and reduce consumed power at electromotor. In the lower running zone of the main propulsion, shafts total consumed power at electromotor will be 7.85% less than real operational power. In the upper running zone of the main propeller shafts, this saving decreases by about 1% and is 6.84%.

The results of this analysis could be useful for a broad audience and for ship owners, compressor operators, and producers.

Author Contributions: Conceptualization, I.P. and J.O.; Data curation, I.P.; Formal analysis, I.P. and V.M.; Funding acquisition, J.O.; Investigation, I.P.; Methodology, V.M. and D.B.; Resources, I.P. and J.O.; Supervision, V.M. and D.B.; Validation, I.P.; Visualization, J.O. and D.B.; Writing—original draft, I.P.; Writing—review & editing, J.O. All authors have read and agreed to the published version of the manuscript.

Funding: This research has been supported by the Croatian Science Foundation under the project IP-2018-01-3739, CEEPUS network CIII-HR-0108, European Regional Development Fund under the grant KK.01.1.1.01.0009 (DATACROSS), University of Rijeka scientific grant uniri-tehnic-18-275-1447, University of Rijeka scientific grant uniri-tehnic-18-18-1146 and University of Rijeka scientific grant uniri-tehnic-18-14.

Conflicts of Interest: The authors declare no conflict of interest.

Nomenclature

Latin Symbols

$\dot{El}$	energy rate loss, kW
ex	specific exergy, kJ/kg
$\dot{Exd}$	exergy destruction rate, kW
g	gravitational acceleration, m/s^2
h	specific enthalpy, kJ/kg
I	amperage, A
$\dot{m}$	mass flow rate, kg/h
P	power, kW
$\dot{Q}$	heat flow rate, kW
$\dot{S}$	entropy rate, kW/K
T	temperature, K
U	voltage, V
$\tilde{v}$	speed, m/s
z	height, m

Greek symbols

η_I energy efficiency, %

η_{II} exergy efficiency, %

φ power factor

Abbreviations

BOG   boil off gas

DFDE   dual fuel diesel electric

EM   electromotor

LNG   liquefied natural gas

rpm   revolutions per minute

Subscript

a   adiabatic

References

1. Kim, K.; Park, K.; Roh, G.; Chun, K. Case Study on Boil-Off Gas (BOG) Minimization for LNG Bunkering Vessel Using Energy Storage System (ESS). *J. Mar. Sci. Eng.* **2019**, *7*, 130. [CrossRef]

2. Cabral, H.; Fonseca, V.; Sousa, T.; Costa Leal, M. Synergistic Effects of Climate Change and Marine Pollution: An Overlooked Interaction in Coastal and Estuarine Areas. *Int. J. Environ. Res. Public Health* **2019**, *16*, 2737. [CrossRef] [PubMed]

3. Koo, J.; Oh, S.-R.; Choi, Y.-U.; Jung, J.-H.; Park, K. Optimization of an Organic Rankine Cycle System for an LNG-Powered Ship. *Energies* **2019**, *12*, 1933. [CrossRef]

4. Mrzljak, V.; Poljak, I.; Medica-Viola, V. Dual fuel consumption and efficiency of marine steam generators for the propulsion of LNG carrier. *Appl. Ther. Eng.* **2017**, *119*, 331–346. [CrossRef]

5. Ahmadi, M.H.; Sadaghiani, M.S.; Pourfayaz, F.; Ghazvini, M.; Mahian, O.; Mehrpooya, M.; Wongwises, S. Energy and Exergy Analyses of a Solid Oxide Fuel Cell-Gas Turbine-Organic Rankine Cycle Power Plant with Liquefied Natural Gas as Heat Sink. *Entropy* **2018**, *20*, 484. [CrossRef]

6. Ali Emadi, M.; Mahmoudimehr, J. Modeling and thermo-economic optimization of a new multi-generation system with geothermal heat source and LNG heat sink. *Energy Convers. Manag.* **2019**, *189*, 153–166. [CrossRef]

7. Sadaghiani, M.S.; Ahmadi, M.H.; Mehrpooya, M.; Pourfayaz, F.; Feidt, M. Process development and thermodynamic analysis of a novel power generation plant driven by geothermal energy with liquefied natural gas as its heat sink. *Appl. Therm. Eng.* **2018**, *133*, 645–658. [CrossRef]

8. Abd Majid, M.A.; Haji Ya, H.; Mamat, O.; Mahadzir, S. Techno Economic Evaluation of Cold Energy from Malaysian Liquefied Natural Gas Regasification Terminals. *Energies* **2019**, *12*, 4475. [CrossRef]

9. Fernández, I.A.; Gómez, M.R.; Gómez, J.R.; Insua, A.B. Review of propulsion systems on LNG carriers. *Renew. Sustain. Energy Rev.* **2017**, *67*, 1395–1411. [CrossRef]

10. Mrzljak, V.; Blecich, P.; Anđelić, N.; Lorencin, I. Energy and Exergy Analyses of Forced Draft Fan for Marine Steam Propulsion System during Load Change. *J. Mar. Sci. Eng.* **2019**, *7*, 381. [CrossRef]

11. Mrzljak, V.; Poljak, I.; Mrakovčić, T. Energy and exergy analysis of the turbo-generators and steam turbine for the main feed water pump drive on LNG carrier. *Energy Convers. Manag.* **2017**, *140*, 307–323. [CrossRef]

12. Chen, Z.; Zhang, F.; Xu, B.; Zhang, Q.; Liu, J. Influence of methane content on a LNG heavy-duty engine with high compression ratio. *Energy* **2017**, *128*, 329–336. [CrossRef]

13. Chen, Z.; Xu, B.; Zhang, F.; Liu, J. Quantitative research on thermodynamic process and efficiency of a LNG heavy-duty engine with high compression ratio and hydrogen enrichment. *Appl. Therm. Eng.* **2017**, *125*, 1103–1113. [CrossRef]

14. Senčić, T.; Mrzljak, V.; Blecich, P.; Bonefačić, I. 2D CFD Simulation of Water Injection Strategies in a Large Marine Engine. *J. Mar. Sci. Eng.* **2019**, *7*, 296. [CrossRef]

15. Li, S.; Wang, B.; Dong, J.; Jiang, Y. Thermodynamic analysis on the process of regasification of LNG and its application in the cold warehouse. *Therm. Sci. Eng. Prog.* **2017**, *4*, 1–10. [CrossRef]

16. Akbari, N. Introducing and 3E (Energy, Exergy, Economic) analysis of an integrated transcritical CO_2 Rankine cycle, Stirling power cycle and LNG regasification process. *Appl. Therm. Eng.* **2018**, *140*, 442–454. [CrossRef]

17. Ghorbani, B.; Mehrpooya, M.; Hamedi, M.H.; Amidpour, M. Exergoeconomic analysis of integrated natural gas liquids (NGL) and liquefied natural gas (LNG) processes. *Appl. Therm. Eng.* **2017**, *113*, 1483–1495. [CrossRef]

18. Yildirim, N.; Genc, S. Energy and exergy analysis of a milk powder production system. *Energy Convers. Manag.* **2017**, *149*, 698–705. [CrossRef]

19. Chen, L.; Luo, J.; Sun, F.; Wu, C. Optimized efficiency axial-flow compressor. *Appl. Energy* **2005**, *81*, 409–419. [CrossRef]

20. Reddy, H.V.; Bisen, V.S.; Rao, H.N.; Dutta, A.; Garud, S.S.; Karimi, I.A.; Farooq, S. Towards energy-efficient LNG terminals: Modeling and simulation of reciprocating compressors. *Comput. Chem. Eng.* **2019**, *128*, 312–321. [CrossRef]

21. Park, J.; In, S.; Ko, J.; Kim, H.; Hong, Y.; Yeom, H.; Park, S. Development and parametric study of a 1kW class free-piston stirling cryocooler (FPSC) driven by a dual opposed linear compressor for LNG (Re)liquefaction. *Int. J. Refrig.* **2019**, *104*, 113–122. [CrossRef]

22. Nawaz, A.; Qyyum, M.A.; Qadeer, K.; Khan, M.S.; Ahmad, A.; Lee, S.; Lee, M. Optimization of mixed fluid cascade LNG process using a multivariate Coggins step-up approach: Overall compression power reduction and exergy loss analysis. *Int. J. Refrig.* **2019**, *104*, 189–200. [CrossRef]

23. Park, H.; Lee, S.; Jeong, J.; Chang, D. Design of the Compressor-Assisted LNG fuel gas supply system. *Energy* **2018**, *158*, 1017–1027. [CrossRef]

24. Wartsila Engines, Wartsila 50DF Product Guide. 2016. Available online: https://cdn.wartsila.com/docs/defa ult-source/product-files/engines/df-engine (accessed on 18 December 2019).

25. Dvornik, J.; Dvornik, S. Dual-Fuel-Electric Propulsion Machinery Concept on LNG Carriers. *Trans. Marit. Sci.* **2014**, *3*, 137–148. [CrossRef]

26. Zhao, B.; Zhou, S.; Feng, J.; Peng, X.; Jia, X. Application of a Fluid–Structure Interaction Model for Analysis of the Thermodynamic Process and Performance of Boil-Off Gas Compressors. *Entropy* **2019**, *21*, 341. [CrossRef]

27. Numaguchi, H.; Satoh, T.; Ishida, T.; Matsumoto, S.; Hino, K.; Iwasaki, T. Japan's first dual-fuel diesel-electric propulsion liquefied natural gas (LNG) carrier. *Mitsubishi Heavy Ind. Tech. Rev.* **2009**, *46*, 1–4. Available online: https://www.mhi.co.jp/technology/review/pdf/e461/e461001.pdf (accessed on 18 December 2019).

28. Shipping, World Trade and the Reduction of CO2 Emissions. United Nations Framework Convention on Climate Change (UNFCCC). International Chamber of Shipping (ICS). Available online: http://www.ics-shipping.org/docs/default-source/resources/environmental-protection/shipping-wor ld-trade-and-the-reduction-of-co2-emissions.pdf?sfvrsn=14 (accessed on 18 December 2019).

29. Nitrogen Oxides (NOX)—Regulation 13. IMO International Maritime Organization. Available online: http://www.imo.org/en/OurWork/Environment/PollutionPrevention/AirPollution/Pages/Nitrogen-oxides-(NOx)-%E2%80%93-Regulation-13.aspx (accessed on 18 December 2019).

30. Rules for Classification—Ships, Part 5—Ship Types, Chapter 7—Liquefied Gas Tankers. DNV-GL, 2017. Available online: https://rules.dnvgl.com/docs/pdf/dnvgl/ru-ship/2017-01/DNVGL-RU-SHIP-Pt5Ch7.pdf (accessed on 15 December 2019).

31. Rules for Classification of Ships—Newbuildings Special Service and Type Additional Class, Part 5, Chapter 5—Liquefied Gas Carriers. DNV. Available online: http://rules.dnvgl.com/docs/pdf/dnv/rulesship/2016-07/t s505.pdf (accessed on 18 December 2019).

32. 2 Stage Gas Compressor CM2 Series. CryoStar. Available online: https://cryostar.com/cryogenic-compresso rs-blowers/multi-stage/ (accessed on 19 December 2019).

33. Morel, K. *Instalation, Operation and Maintenance Manual for Fuel Gas Compressor Type CM2-20*; Internal Ship Document; Cryostar: Dublin, Ireland, 2016; p. 174.

34. Gauge Pressure Transmitter EJX510A. Yokogawa. Available online: https://web-material3.yokogawa.com/G S01C25F01-01EN.pdf (accessed on 20 December 2019).

35. Thermowell Fabricated Jumo, PT-100. Available online: https://www.jumo.net (accessed on 20 December 2019).

36. Micro-motion Model 1700. Emerson Process Management. Available online: https://www.emerson.com/docu ments/automation/configuration-use-manual-transmitters-analog-outputs-model-1700-micro-motion-en-624 54.pdf (accessed on 19 December 2019).

37. Protection Relay Vamp 200 Series. Schneider Electric. Available online: https://www.se.com/ww/en/product -range/62048-vamp-200-series/?subNodeId=12146923337en_WW (accessed on 19 December 2019).

38. SEC Power Meter. Shoyo Engineering Co., Ltd. Available online: https://www.shoyo-e.co.jp/english/shaft-horsepower-meter/se203c/ (accessed on 17 December 2019).

39. Sabhi, S. *Electric Propulsion Remote Control*; General Electric: Schenectady, NY, USA, 2017; pp. 13–67.

40. Variable Speed Drives. Carbon Trust. Available online: https://www.carbontrust.com/media/13063/ctg070_variable_speed_drives.pdf (accessed on 18 December 2019).

41. ABB. Technical Guide No. 4, Guide to Variable Speed Drives. Available online: https://library.e.abb.com/public/d3c711ec2acddb18c125788f002cf5da/ABB_Technical_guide_No_4_REVC.pdf (accessed on 18 December 2019).

42. Variable Frequency Drive—Operation and Application of Variable Frequency Drive (vfd) Technology. Carrier Corporation Syracuse. Available online: https://www.utcccs-cdn.com/ (accessed on 18 December 2019).

43. Poljak, I. *Certificate of quality, Sampling and Analysis, Analitic Results*; SGS: Baltimore, MD, USA, 2018.

44. Dincer, I.; Rosen, M. *Exergy, Energy, Environment and Sustainable Development*; Elsevier: Oxford, UK, 2007; pp. 23–30.

45. Poljak, I.; Orović, J.; Mrzljak, V. Energy and exergy analysis of the condensate pump during internal leakage from the marine steam propulsion system. *Sci. J. Mar. Res.* **2018**, *32*, 268–280. [CrossRef]

46. Kanoglu, M.; Cengel, Y.A.; Dincer, I. *Efficiency Evaluation of Energy Systems*; Springer: Berlin, Germany, 2012; pp. 9–40. [CrossRef]

47. Mrzljak, V.; Poljak, I. Energy Analysis of Main Propulsion Steam Turbine from Conventional LNG Carrier at Three Different Loads. *Int. J. Mar. Sci. Technol.* **2019**, *66*, 10–18. [CrossRef]

48. Moran, M.J.; Shapiro, H.N.; Boettner, D.D.; Bailey, M.B. *Fundamentals of Engineering Thermodynamics*, 8th ed.; John Wiley and Sons: Hoboken, NJ, USA, 2014; pp. 190–191, 337–339, 403.

49. Chen, W.K. *The Electrical Engineering Handbook*; Elsevier, Academic Press: Oxford, UK, 2005; p. 810.

50. Maxfield, C.; Bird, J.; Laughton, M.A.; Bolton, W.; Leven, A.; Schmitt, R.; Sueker, K.; Williams, T.; Tooley, M.; Moura, L.; et al. *Electrical Engineering: Know It All*; Elsevier: Oxford, UK, 2008; pp. 6–7.

51. Ahmadi, G.R.; Toghraie, D. Energy and exergy analysis of Montazeri Steam Power Plant in Iran. *Renew. Sustain. Energy Rev.* **2016**, *56*, 454–463. [CrossRef]

52. Szargut, J. *Exergy Method—Technical and Ecological Applications*; WIT Press: Southampton, UK, 2004; pp. 15–42.

53. Orović, J.; Mrzljak, V.; Poljak, I. Efficiency and Losses Analysis of Steam Air Heater from Marine Steam Propulsion Plant. *Energies* **2018**, *11*, 3019. [CrossRef]

54. Nazari, N.; Heidarnejad, P.; Porkhial, S. Multi-objective optimization of a combined steam-organic Rankine cycle based on exergy and exergo-economic analysis for waste heat recovery application. *Energy Convers. Manag.* **2016**, *127*, 366–379. [CrossRef]

55. Hepbasli, A. A key review on exergetic analysis and assessment of renewable energy resources for a sustainable future. *Renew. Sustain. Energy Rev.* **2008**, *12*, 593–661. [CrossRef]

56. Mrzljak, V.; Senčić, T.; Žarković, B. Turbogenerator Steam Turbine Variation in Developed Power: Analysis of Exergy Efficiency and Exergy Destruction Change. *Model. Simul. Eng.* **2018**, *2018*, 2945325. [CrossRef]

57. Mohtaram, S.; Chen, W.; Zargar, T.; Lin, J. Energy-exergy analysis of compressor pressure ratio effects on thermodynamic performance of ammonia water combined cycle. *Energy Convers. Manag.* **2017**, *134*, 77–87. [CrossRef]

58. Dincer, I.; Midilli, A.; Kucuk, H. *Progress in Exergy, Energy and the Environment*; Springer: Basel, Switzerland, 2014; pp. 105–115.

59. Mrzljak, V.; Poljak, I.; Prpić-Oršić, J. Exergy analysis of the main propulsion steam turbine from marine propulsion plant. *Shipbuild. Theory Pract. Nav. Archit. Mar. Eng. Ocean Eng.* **2019**, *70*, 59–77. [CrossRef]

60. Lorencin, I.; Anđelić, N.; Mrzljak, V.; Car, Z. Exergy analysis of marine steam turbine labyrinth (gland) seals. *Sci. J. Mar. Res.* **2019**, *33*, 76–83. [CrossRef]

61. Adibhatla, S.; Kaushik, S.C. Energy and exergy analysis of a super critical thermal power plant at various load conditions under constant and pure sliding pressure operation. *Appl. Ther. Eng.* **2014**, *73*, 51–65. [CrossRef]

62. Ameri, M.; Ahmadi, P.; Hamidi, A. Energy, exergy and exergoeconomic analysis of a steam power plant: A case study. *Int. J. Energy Res.* **2009**, *33*, 499–512. [CrossRef]

Article

Second Law Analysis for the Experimental Performances of a Cold Heat Exchanger of a Stirling Refrigeration Machine

Steve Djetel-Gothe, François Lanzetta * and Sylvie Bégot

FEMTO-ST, Energy Department, Univ. Bourgogne Franche-Comté, CNRS Parc technologique,
2 avenue Jean Moulin, 90000 Belfort, France; steve.djetel@femto-st.fr (S.D.-G.); sylvie.begot@univ-fcomte.fr (S.B.)
* Correspondence: francois.lanzetta@univ-fcomte.fr; Tel.: +33-3-84-57-82-24

Received: 23 December 2019; Accepted: 11 February 2020; Published: 14 February 2020

Abstract: The second law of thermodynamics is applied to evaluate the influence of entropy generation on the performances of a cold heat exchanger of an experimental Stirling refrigeration machine by means of three factors: the entropy generation rate N_S, the irreversibility distribution ratio ϕ and the Bejan number $Be|_{N_S}$ based on a dimensionless entropy ratio that we introduced. These factors are investigated as functions of characteristic dimensions of the heat exchanger (hydraulic diameter and length), coolant mass flow and cold gas temperature. We have demonstrated the role of these factors on the thermal and fluid friction irreversibilities. The conclusions are derived from the behavior of the entropy generation factors concerning the heat transfer and fluid friction characteristics of a double-pipe type heat exchanger crossed by a coolant liquid (55/45 by mass ethylene glycol/water mixture) in the temperature range 240 K < T_C < 300 K. The mathematical model of entropy generation includes experimental measurements of pressures, temperatures and coolant mass flow, and the characteristic dimensions of the heat exchanger. A large characteristic length and small hydraulic diameter generate large entropy production, especially at a low mean temperature, because the high value of the coolant liquid viscosity increases the fluid frictions. The model and experiments showed the dominance of heat transfer over viscous friction in the cold heat exchanger and $Be|_{N_S} \to 1$ and $\phi \to 0$ for mass flow rates $\dot{m} \to 0.1$ kg.s^{-1}.

Keywords: Stirling cycle; refrigerator; heat exchanger; second law; entropy production

1. Introduction

Refrigeration plays a major role in many different sectors, ranging from food, air conditioning, healthcare, industry and, especially energy. Nowadays, the number of refrigeration air-conditioning and heat pump systems in operation worldwide is roughly 5 billion, if we consider 2.6 billion air-conditioning stationary and mobile units and 2 billion domestic refrigerators and freezers [1]. The global electricity demand for refrigeration, including air conditioning, could more than double by 2050 [2]. The Kyoto Protocol was adopted in 1997 and entered into force on 16 February 2005. It aimed to prevent global warming due to the use of HCFC/CFC refrigerants and the objective is to stabilise greenhouse gas concentrations at a level that would prevent dangerous anthropogenic (human-induced) interference with the climate system. It can be seen that developing a cooling system without using HCFC/CFC working fluid is unavoidable in the future. Stirling coolers are used as commercial cryocoolers in the cryogenic field, military applications, liquid air production plants, cooling electronics, carbon capture and domestic applications [3–6].

Thus, in order to decrease electricity consumption and to prevent global warning, it is necessary to optimize these cooling machines and systems, particularly the heat exchangers. These are systems to transfer heat between two or more fluids at different temperatures and pressures, separated by a heat-transfer surface. The optimal conversion of heat energy involves a heat exchanger in order to minimize the energy consumption and costs regarding several criteria, such as materials, geometries, flow rates, flow arrangements, operating temperatures and pressures, and transient or steady-state operation [7–10]. In the field of refrigeration, the objective is to extract heat from different products in gas, liquid or solid state with minimum energy cost and maximum efficiency.

This paper deals with the optimization of a heat exchanger (the freezer) of a refrigerating machine, used in a Stirling machine working at low and moderate temperatures (between -100 and 0 °C). This machine is a member of the cryocooler's technology family (Figure 1). Generally, the closed cycle of the Stirling machine concerns engines, coolers and heat pumps [11,12]. The thermodynamic cycle is the same in each case except that the process direction is reversed. In its theoretical refrigerating cycle (Figures 2–4), the working fluid is compressed at the highest constant temperature T_H (1–2) and Q_H is the corresponding heat rejected to the environment through the cooler exchanger.

The fluid is cooled at constant volume (2–3) by the heat Q_R rejected from the gas to the regenerator. Then, the expansion work (3–4) takes place at the lowest constant temperature T_C and the external heat Q_C is extracted from the surroundings and supplied to the gas through the freezer exchanger. Finally, the fluid is heated from the temperature T_C to T_H by the corresponding heat Q_R stored in the regenerator during the process at constant volume (4–1). Both isochoric processes take place in a porous heat exchanger called the regenerator [4,5], whose efficiency is a key point of Stirling machine performances.

In refrigeration operation, heat is then rejected to the hot sink during the compression (1–2) and provided to the cold source in the compression stage (3–4). In a Stirling machine, these exchangers play a crucial role in the optimization of the performances. The objective of this paper is to optimize the performance of these exchangers by means of the second law of analysis and to study the behavior of the entropy generation. This approach focuses on the cold source exchanger (the freezer) at constant temperature T_C (Figure 5), considering that the same approach could concern the exchanger of the hot sink at temperature T_H.

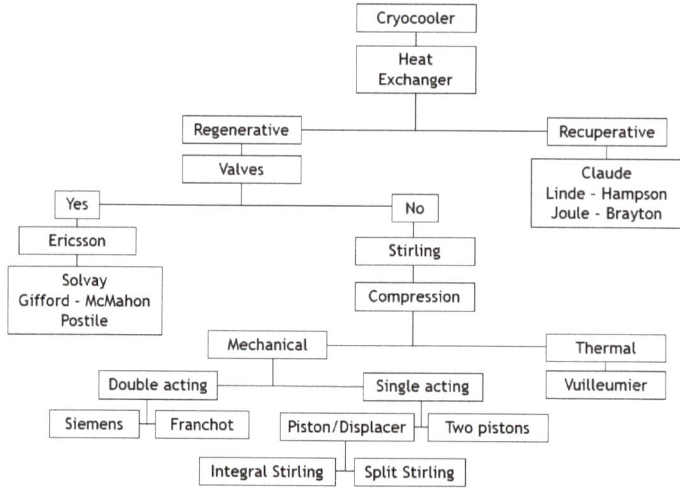

Figure 1. Classification of cryocoolers [12].

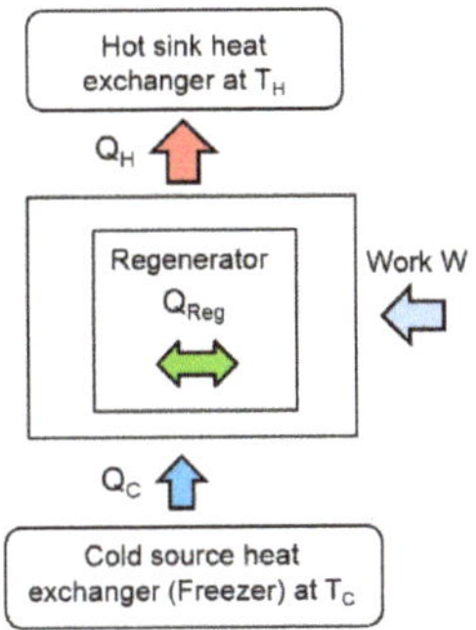

Figure 2. Schematic diagram of an ideal Stirling refrigerator.

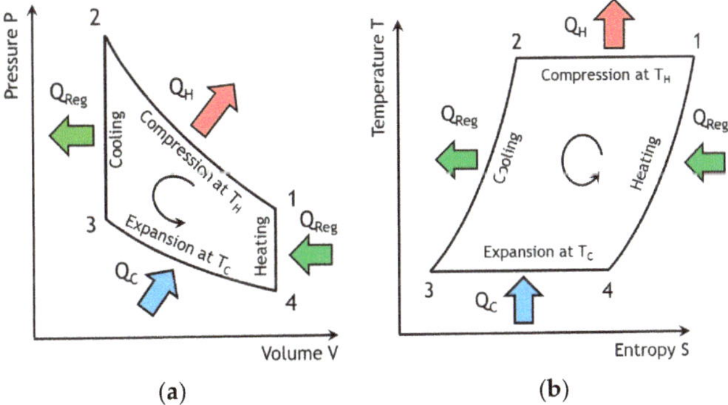

Figure 3. Ideal Stirling refrigeration cycle. (**a**) Pressure-Volume diagram; (**b**) Temperature-Entropy diagram.

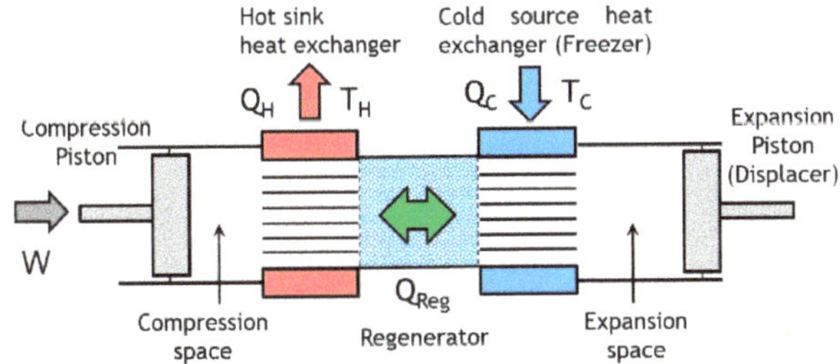

Figure 4. Stirling refrigerator for an alpha architecture.

Entropy generation analysis is considered as a measure of sustainability because a process with a lower entropy generation rate is more sustainable and able to realize energy conversion more efficiently. Nowadays, new indicators based on an engineering approach of irreversibility are used to evaluate both the technological level and the environmental impact of the production process and the socio-economic conditions [13–15]. Heat and mass transfer processes and thermal systems like heat exchangers have been studied and optimized in order to minimize the irreversibilities using the second law of thermodynamics in fundamental works published since the 1980s [16–19]. The second law of thermodynamics method is applied to general convective heat transfer flows in ducts and it defines the entropy rate contributed by heat transfer irreversibilities and fluid flow frictions. Convective heat transfer through ducts is analyzed with constant heat flux or constant temperature [20–24].

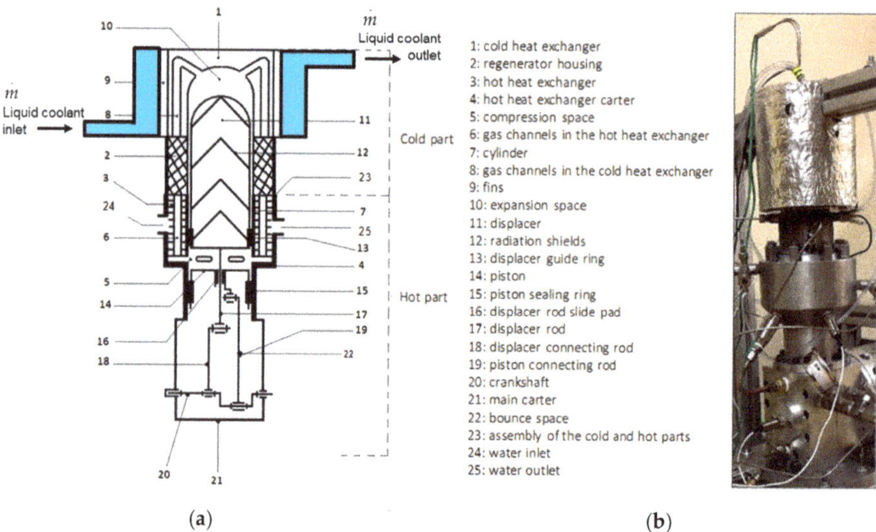

Figure 5. Stirling refrigerating machine [25]. (**a**) Profile of the mechanical configuration; (**b**) The cold heat exchanger is thermally insulated (white material on the photo).

The main objective of this work is to determine the optimum duct or heat exchanger geometry which minimize losses for a range of Reynold numbers, hydraulic diameters, and fins geometry (thickness, spacing). We describe the experimental setup based on a Beta Stirling refrigerator, specifically focusing on the cold heat exchanger. Then, we will focus our study on the second law analysis of the experimental performances of the cold heat exchanger. The model and optimization present complementary criteria and alternative ratios, like the dimensionless entropy ratio, irreversibility distribution ratio and Bejan number, that we used in this study to determine the performances of a cold heat exchanger of a Stirling refrigerator.

2. Experimental Setup

The Stirling refrigerator in the Beta configuration (Figure 5) is a single-cylinder machine. It consists of two pistons, a working piston and a displacer piston, a compression volume, an expansion volume, a volume occupied by a regenerator, and two heat exchangers. The working piston is used to compress and expand the fluid. The displacer piston function is to displace the gas between the hot and cold parts of the machine. The function of the regenerator is to store the heat of the fluid during one of the isochoric transformations of the cycle and to restore it in the other one. The heat exchanger of the hot part of the machine is water-cooled. The lower part of the machine, called the bounce space, contains the connecting rods and the crankshaft.

The main design parameters of the machine are: mean pressure between 15 and 20 bar, Nitrogen as working gas, stainless steel for the cold head and aluminum for the radiator at the hot sink (Table 1). The experimental cooling capacity is 550 W at 273 K and 280 W at 230 K.

Table 1. Mechanical characteristics of the Stirling refrigerator.

Characteristics	Values
Cooling capacity at 230 K/273 K	280 W/550 W
Cold end temperature	190 K < T_C < 273 K
Hot sink temperature	T_H = 300 K
Working gas	Nitrogen
Mean Pressure	15 bar < P < 18 bar
Stirling refrigerator overall dimensions Width × Height × Depth	19 × 46 × 18 cm
Power piston diameter	D = 60 mm
Power piston stroke	C = 40 mm
Compression swept volume	V_{swc} = 120 cm^3
Regenerator	Stainless steel wire mesh
Rotational speed	35 rad.s^{-1} < ω < 80 rad.s^{-1}

3. Mathematical Model

In a heat exchanger, real processes present irreversibilities caused by losses due to fluid friction, heat exchange across a finite temperature difference and heat exchange with the environment [26]. Second law analysis is a method developed to design systems on minimum entropy production caused by these losses [16,27,28]. The second law equation describes the irreversibility of the process, in terms of entropy generation rate $\dot{S}_{gen}$ within the heat exchanger between inlet and outlet ducts of a system boundary.

$$\dot{S}_{gen} = \frac{\partial S}{\partial t} - \frac{\dot{Q}}{T} - \sum_{in} \dot{m}_{in}s + \sum_{out} \dot{m}_{out}s \geq 0 \tag{1}$$

For steady operation ($\partial S / \partial t = 0$), consider an arbitrary flow passage of length dx with a finite temperature difference $\Delta T = (T_w - T)$ between the wall temperature T_w and the bulk temperature T of the fluid, the rate of entropy generation per unit length is

$$\dot{S}'_{gen} = \frac{\dot{Q}'\Delta T}{T^2(1+\tau)} + \frac{\dot{m}}{\rho T}\left(-\frac{dP}{dx}\right) \tag{2}$$

where $\tau = \Delta T/T$ a dimensionless temperature difference, $\dot{Q}'$ the wall heat transfer per unit length, $\dot{m}$ the mass flow rate and $(-dP/dx)$ the longitudinal pressure gradient.

The quantity $\frac{\dot{Q}'\Delta T}{T^2(1+\tau)}$ corresponds to the entropy generation rate accounting for the heat transfer irreversibility ($\dot{S}'_{gen,\Delta T}$) and the term $\frac{\dot{m}}{\rho T}\left(-\frac{dP}{dx}\right)$ corresponds to the entropy generation rate for the fluid friction irreversibility ($\dot{S}'_{gen,\Delta P}$). Based on these quantities, Bejan defined the irreversibility distribution ratio ϕ by [29]:

$$\phi = \frac{\dot{S}'_{gen,\Delta P}}{\dot{S}'_{gen,\Delta T}} = \frac{\frac{\dot{m}}{\rho T}\left(-\frac{dP}{dx}\right)}{\frac{\dot{Q}'\Delta T}{T^2(1+\tau)}} \tag{3}$$

The entropy generation irreversibility can be described by the Bejan number Be as the ratio between the heat transfer irreversibility and the total irreversibility due to heat transfer and fluid flow [30,31]:

$$Be = \frac{\dot{S}'_{gen,\Delta T}}{\dot{S}'_{gen,\Delta T} + \dot{S}'_{gen,\Delta P}} = \frac{1}{1+\phi} \tag{4}$$

The heat transfer dominates when $Be \rightarrow 1$ (or $\phi \rightarrow 0$) and the fluid friction dominates when $Be \rightarrow 0$ (or $\phi \rightarrow \infty$). The ideal equilibrium between these two irreversibilities is reached for $Be = 1/2$ or ($\phi = 1$).

Assuming the refrigerating fluid to be incompressible, Equation (2) can be written as:

$$\dot{S}'_{gen} = \dot{m}c_p\underbrace{\frac{dT}{dx}\frac{\Delta T}{T^2(1+\tau)}}_{\dot{S}'_{gen,\Delta T}} + \underbrace{\frac{\dot{m}}{\rho T}\left(-\frac{dP}{dx}\right)}_{\dot{S}'_{gen,\Delta P}} \tag{5}$$

Equation (6) represents the entropy generation per unit length due to heat transfer across finite temperature difference and to fluid friction, respectively. If we consider that the dimensionless temperature difference $\tau = \Delta T/T \ll 1$, the following equation is obtained:

$$\dot{S}'_{gen} = \dot{m}c_p\frac{dT}{dx}\frac{\Delta T}{T^2} + \frac{\dot{m}}{\rho T}\left(-\frac{dP}{dx}\right) \tag{6}$$

Derived from Figure 5, the Figure 6 presents the control volume of the Stirling cold heat exchanger for energy and entropy analysis.

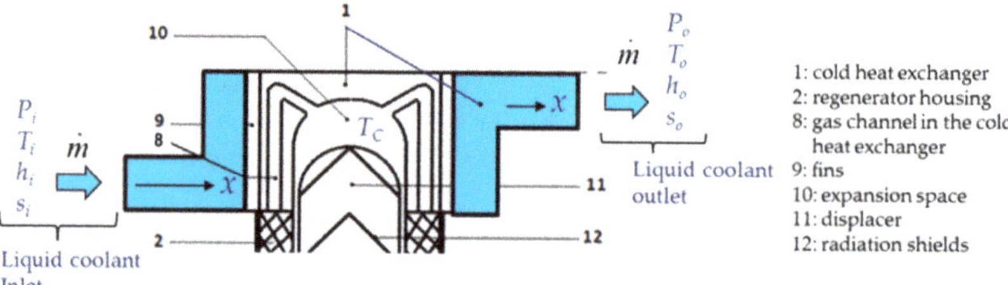

Figure 6. Control volume of the cold heat exchanger.

The cold heat exchanger is a double-pipe type heat exchanger made up of two concentric circular tubes (Label 1 in Figure 6). The incompressible coolant liquid flows continuously, with a mass flow $\dot{m}$, through the annular passage under the temperature $(T_i - T_o)$ and pressure $(P_i - P_o)$ gradients, respectively. The other fluid is the working gas (Nitrogen) of the Stirling refrigerator and it flows alternatively (corresponding to the rotational frequency of the machine) at constant temperature T_C in the inner tube corresponding to the cylinder of the machine (Labels 8 and 10). The heat flux $\dot{Q}$ is extracted from the surface and maintained at the constant temperature T_C along the heat exchanger surface. This temperature is measured in the expansion volume of the refrigerator and depends on different parameters like thermal load, pressure of the gas, rotational speed [12,32]. The heat exchanger is considered as a duct at constant temperature T_C with a hydraulic diameter D_h and a total length L corresponding to mean distance between the inlet and the outlet of the heat exchanger.

Equation (7) is integrated along the length, and the expression of the entropy generation $\dot{S}_{gen}$ becomes [21,22,33]:

$$\dot{S}_{gen} = \dot{m}c_p(T_o - T_i)\frac{(T_c - T_o)}{T_iT_o} + \frac{32\dot{m}^3f}{\rho^2\pi^2D_h^5}\frac{L}{T_C} \tag{7}$$

where $-dP/dx = 2f\rho u_m^2/D_h$ and $u_m = 4\dot{m}/\left(\rho\pi D_h^2\right)$.

The hydraulic diameter approach is a simple dimensional parameter method to calculate the heat transfer and the pressure drop in the heat exchanger. The hydraulic diameter D_h of the cold heat exchanger is defined [34].

$$D_h = 4\left(\frac{\text{minimum flow area}}{\text{frontal area}}\right)\left(\frac{\text{total volume}}{\text{total surface area}}\right) \tag{8}$$

The rate of entropy generation is represented as the dimensionless form N_S [29]:

$$N_S = \frac{\dot{S}_{gen}}{\dot{m}c_p} \tag{9}$$

where $\left(\dot{m}c_p\right)$ is the heat capacity rate within the heat exchanger. In this work, the heat capacity rate is the product of the mass flow and specific heat capacity rate of the incompressible fluid.

The fluid flow is considered to be fully developed inside the exchanger. The friction factor f is given for a well-known correlation [35–37]. The complex flow field generated by the internal geometry of the heat exchanger influences the pressure gradient required to drive the flow. The friction factor f is defined

$$f = aRe^b \tag{10}$$

with the Reynolds number $Re = \rho u_m D_h / \mu$. The two constants a and b depend on the fluid properties, the flow regime and the geometry constraints of the heat exchanger. These two coefficients were identified from data reduction experimental results [38] and pressure drop data were correlated to $\pm 10\%$ for $Re < 2000$ (laminar flow condition).

$$f = 34.9 \, Re^{-0.775} \tag{11}$$

The average of the inlet and outlet bulk temperatures is used to calculate the physical parameters (density, viscosity and specific heat capacity) of the coolant liquid (55/45 by mass ethylene glycol/water mixture) in the temperature range 240 K $< T <$ 300 K (Table 2 and Figure 7). Ethylene glycol lowers the specific heat capacity of water mixtures relative to pure water.

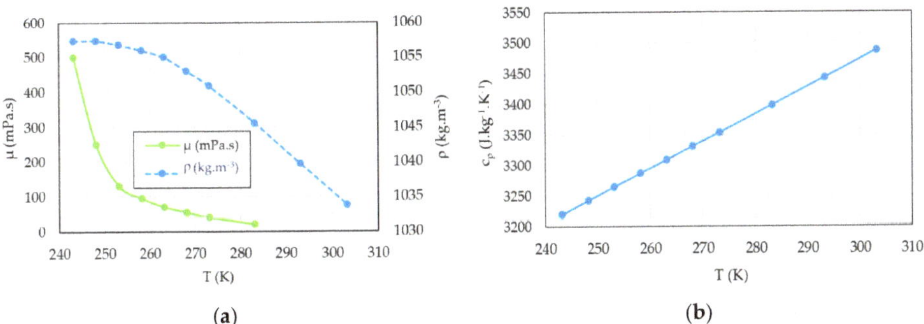

Figure 7. Physical properties of the coolant liquid (55/45 by mass ethylene glycol/water mixture) in the temperature range 240 K $< T <$ 300 K [39] (a) Density ρ (kg.m^{-3}) and dynamic viscosity μ (mPa.s); (b) Specific heat capacity c_p (J.kg^{-1}.K^{-1}).

Considering Equation (10), we get:

$$N_S = \underbrace{(T_0 - T_i)\frac{(T_C - T_0)}{T_i T_0}}_{N_{S,\Delta T}} + \underbrace{\frac{32\dot{m}^2 f}{c_p \rho^2 \pi^2 D_h^5}\frac{L}{T_C}}_{N_{S,\Delta P}} \tag{12}$$

The first term corresponds to dimensionless rate of entropy production due to the thermal irreversibilities ($N_{S,\Delta T}$), and the second one, ($N_{S,\Delta P}$), to the pressure irreversibilities.

From Equation (13), we introduce a Bejan number $Be|_{N_S}$ based on a dimensionless entropy ratio:

$$Be|_{N_S} \;=\; \frac{N_{S,\Delta T}}{N_{S,\Delta T} + N_{S,\Delta P}} \;=\; \frac{(T_0 - T_i)\frac{(T_C - T_0)}{T_i T_0}}{(T_0 - T_i)\frac{(T_C - T_0)}{T_i T_0} + \frac{32\dot{m}^2 f}{c_p \rho^2 \pi^2 D_h^5}\frac{L}{T_C}} \tag{13}$$

Table 2. Physical properties of of the coolant liquid (55/45 by mass ethylene glycol/water mixture) in the temperature range 240 K < T < 300 K [39].

Temperature		Thermal Conductivity	Density	Specific Heat Capacity	Dynamic Viscosity	Prandtl Number
T (°C)	T (K)	λ (W.m^{-1}.K^{-1})	ρ (kg.m^{-3})	c_p (J.kg^{-1}.K^{-1})	μ (mPa.s)	Pr
−30	243.15	0.3300	1057.40	3220.2	500	4879.1
−25	248.15	0.3360	1057.40	3242.5	250	2412.6
−20	253.15	0.3377	1056.80	3264.8	130	1256.8
−15	258.15	0.3406	1056.00	3287.1	95	916.8
−10	263.15	0.3406	1055.00	3309.4	70	680.1
−5	268.15	0.3406	1053.00	3331.7	55	538.0
0	273.15	0.3440	1050.90	3354.0	40	390.0
5	278.15	0.3480	1048.25	3376.3	30	291.1
10	283.15	0.3507	1045.60	3398.6	20	193.8
15	288.15	0.3536	1042.65	3420.9		
20	293.15		1039.70	3443.2		
30	303.15		1033.80	3487.8		

4. Results and Discussion

Both fluid friction and heat transfer contribute to the rate of entropy generation represented by Equations (4), (5), (10) and (13). Entropy generation is investigated with the effects of different parameters concerning the cold heat exchanger (Length L, hydraulic diameter D_h, mass flow $\dot{m}$, temperature T_C of the gas inside the cold volume of the Stirling refrigerator). Before examining the dependence of these parameters on entropy generation, it is necessary to discuss the following observations and assumptions. Measurements are performed in the cold heat exchanger (the freezer) of the Stirling refrigerator and concern the temperatures and pressures of the working gas inside the expansion volume corresponding to the internal cold heat exchanger, mass flow and pressures of the coolant liquid (mix water/ethylene–glycol) and rotational speed of the refrigerator (Figure 6).

In this section, we used constant values of inlet and outlet temperatures fixed at T_i = 300 K and T_o = 260 K, respectively. The temperature T_o corresponds to a constraint we need for an industrial process not described in this article. The hydraulic diameter of the cold heat exchanger is D_h = 0.015 m and its characteristic length is L = 0.10 m. For the coolant liquid, the variation of viscosity with temperature is responsible for most of the property effects. The coolant fluid flow $\dot{m}$ vary in the range (0.003–0.010 kg.s^{-1}). In our study, we performed experiments with two mass flows rates of coolant liquid $\dot{m}$ = 0.003 kg.s^{-1} and $\dot{m}$ = 0.004 kg.s^{-1}, respectively. At low temperatures, the viscosity increases significantly and between 243 K and 263 K the dynamic viscosity increases in a ratio of around 7 (Table 2) and has the direct effect of increasing the fluid frictions. We used the film temperature based on the wall and mean bulk temperatures of the liquid flow rate to calculate its properties. We performed the measurements when the machine was operating in steady state at a fixed rotational speed ω with 35 rad.s^{-1} < ω < 80 rad.s^{-1} and for a given working gas temperature T_C with 190 K < T_C < 273 K. At each extremity of the exchanger (inlet and outlet on Figure 6), we collected the characteristic data of the coolant fluid: the refrigerant mass flow rate $\dot{m}$, the inlet and outlet pressures P_i and P_o and their corresponding temperatures T_i and T_o, respectively. Then, we introduced the experimental data in the thermodynamic model concerning the temperatures, the pressures, the mass flows, the hydraulic diameter, and the mean length L of the cold heat exchanger. We finally used the model to extrapolate the performances of the refrigerator in terms of irreversibility distribution ratio ϕ (Equation (3)), entropy generation rate N_S (Equations (10) and (13)) and Bejan numbers $Be|_{N_S}$ (Equations (5) and (14)).

4.1. Effect of Characteristic Dimensions (Length L and Hydraulic Diameter D_h)

Figures 8–10 present the entropy generation parameters (N_S, $Be|_{N_S}$ and φ) as functions of length (L) and hydraulic diameter (D_h) for fixed values of cold gas temperature (T_C). The entropy generation number (N_S) increases with the length L of the heat exchanger because of the increase in the fluid friction and pressure drop. For a given length, N_S (Figure 8) increases when the temperature T_C decreases because the heat transfer and the entropy production increase. For each mass flow rate $\dot{m}$ of the cooling liquid, we measured the temperature T_C of the cold gas inside the expansion space of the refrigerator. For $\dot{m} = 0.003$ kg.s^{-1}, this temperature is $T_{Cexp} = 242$ K and it corresponds to the entropy generation number $N_S \approx 0.010$, while the theoretical model provides a higher value $N_S \approx 0.011$ (+10%) and a lower temperature $T_C = 240$ K. When the flow rate increases ($\dot{m} = 0.004$ kg.s^{-1}), the pressure drop also logically increases, causing the decrease in N_S.

The irreversibility distribution ratio φ increases with the length L of the heat exchanger (Figure 9). At given length L, the smallest values of φ correspond to the case for which the heat transfer dominates. We observe at L = 0.10 m (Figure 9), an increase in φ as the mass flow rate $\dot{m}$ increases because the fluid frictions dominate, as was observed in the literature [3,22,40].

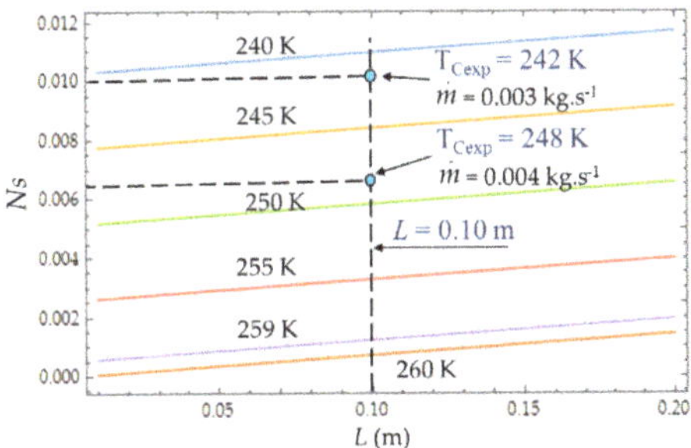

Figure 8. Variations of entropy generation rate N_S as function of cold heat exchange L for different temperatures T_C. Experimental results for $D_h = 0.015$ m, $L = 0.010$ m, $= 0.003$ kg.s^{-1} and $\dot{m} = 0.004$ kg.s^{-1}.

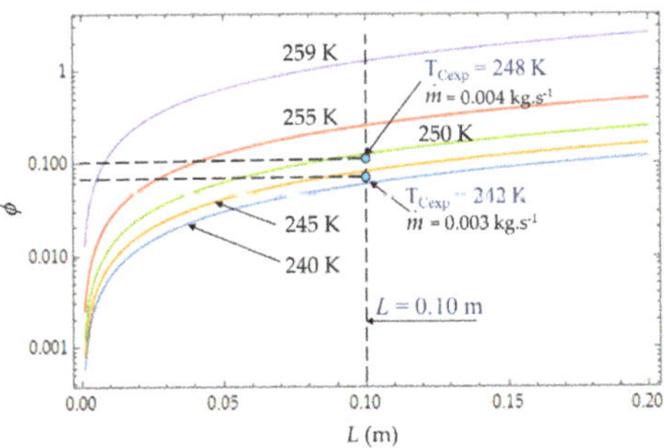

Figure 9. Variation of the irreversibility distribution ratio φ as function of cold heat exchange L for different temperatures T_C. Experimental results for $D_h = 0.015$ m, $L = 0.10$ m, $\dot{m} = 0.003$ kg.s^{-1} and $\dot{m} = 0.004$ kg.s^{-1}.

Figure 10 plots the evolution of $Be|_{N_S}$ as a function of cold heat exchange length L at different temperatures T_C. The limit $Be|_{N_S} \rightarrow 1$ when the heat transfer dominates. At given mass flow rate $\dot{m}$, for small values of hydraulic diameter ($D_h < 5$ mm), the effect of fluid friction is more effective and $Be|_{N_S}$ varies with great sensitivity to temperature T_C. For higher values of hydraulic diameter, $D_h > 5$ mm, the heat transfer dominates fluid frictions regardless of the temperature T_C of the cold volume and, consequently, $\phi \rightarrow 0$ (Figure 11). The effects of the variation in the mass flows $\dot{m} = 0.003$ kg.s^{-1} and $\dot{m} = 0.004$ kg.s^{-1} are of the same amplitude in our experiments and show the dominance of the heat transfer on the fluid frictions ($Be|_{N_S} \rightarrow 1$ and $\phi \rightarrow 0$).

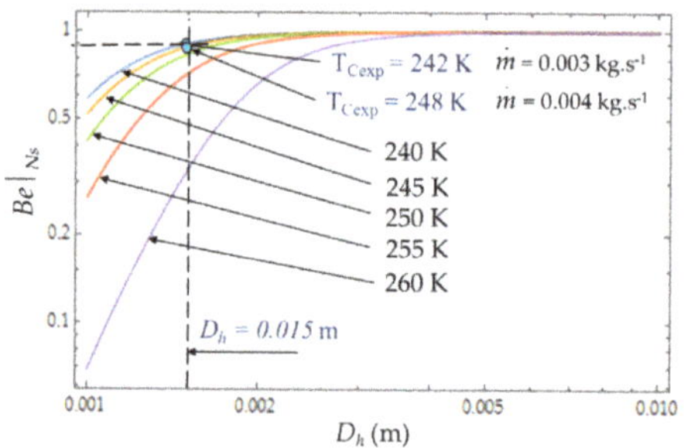

Figure 10. Variations of Bejan number $Be|_{N_S}$ as function of hydraulic diameter D_h for different temperatures T_C. Experimental results for $D_h = 0.015$ m, $L = 0.1$ m, $\dot{m} = 0.003$ kg.s^{-1} and $\dot{m} = 0.004$ kg.s^{-1}.

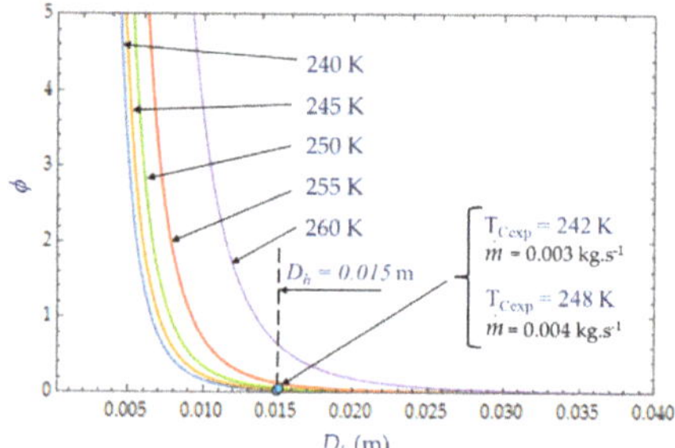

Figure 11. Variation of the irreversibility distribution ratio φ as function of hydraulic diameter D_h for different temperatures T_C. Experimental results for $D_h = 0.015$ m, $L = 0.1$ m, $\dot{m} = 0.003$ kg.s^{-1} and $\dot{m} = 0.004$ kg.s^{-1}.

The heat transfer impacts the entropy generation rate N_S, which logically decreases with the temperature T_C, and the hydraulic diameter D_h, corresponding to the dominance of the heat transfer irresversibilities over fluid losses (Figure 12). That is what we observed with the two experimental temperatures $T_{Cexp} = 242$ K and $T_{Cexp} = 248$ K corresponding to the two mass flow rates $\dot{m} = 0.003$ kg.s^{-1} and $\dot{m} = 0.004$ kg.s^{-1}.

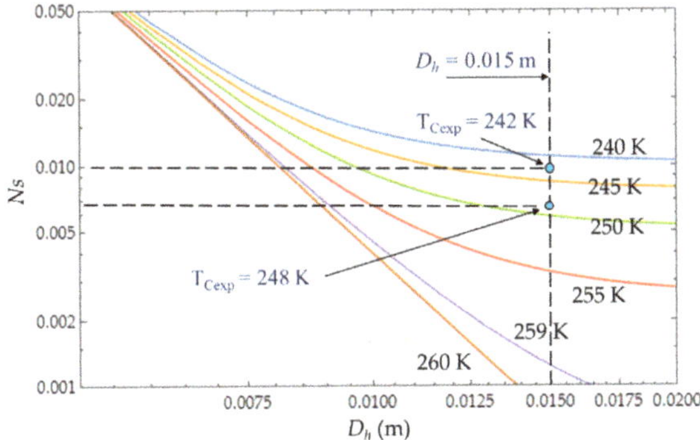

Figure 12. Variations of entropy generation rate N_S as function of hydraulic diameter D_h for different temperatures T_C. Experimental results for $D_h = 0.015$ m, $L = 0.1$ m, $\dot{m} = 0.003$ kg.s^{-1} and $\dot{m} = 0.004$ kg.s^{-1}.

The objective is to reduce the degree of irreversibility of the refrigerator and to obtain small values of N_S [16,22,33,41]. This objective depends on the parameters temperature of the cold volume and hydraulic diameter. We observe that the properties of the refrigerant play an important role. When the temperature of the fluid is too low, the viscosity of the fluid and the pressure drops increase and N_S increases as well, which reduces the performance of the refrigerator. For large values of D_h, the entropy generated by the fluid losses is no longer affected by the flow friction losses. Ns tends towards a constant (in our case, this phenomenon is quite obvious for the temperatures 240 K < Tc < 250 K) and the irreversibility distribution ratio φ tends to zero (Figure 11). The entropy generation rate N_S depends mainly on the part of the entropy due to heat transfers and then decreases as a function of temperature T_C and hydraulic diameter D_h across a trade-off between fluid flow irreversibilities and heat transfer.

4.2. Effect of Coolant Mass Flow $\dot{m}$

As shown in Equations (4), (8) and (13), it appears that the greater the mass flow $\dot{m}$, the higher the fluid frictions and pressure drop. As a result, at a fixed mass flow, both entropy generation rate N_S and irreversibility distribution ratio ϕ increase with the heat transfer (Figure 13). Effectively, at low temperatures ($T_C = 240$ K), the heat transfer from the coolant at mean bulk temperature $T_{mb} = (T_i + T_o)/2$ to the gas is higher than at high temperatures ($T_C = 260$ K) and Ns presents highest variations. This observed for the two experimental temperatures $T_{Cexp} = 242$ K and $T_{Cexp} = 248$ K corresponding to the mass flow rates $\dot{m} = 0.003$ kg.s^{-1} and $\dot{m} = 0.004$ kg.s^{-1}, respectively. Consequently, at a low temperature ($T_C = 240$ K) and small cooling flow rate, the irreversibility distribution ϕ and Bejan number $Be|_{Ns}$ confirm that heat transfer dominates fluid frictions (Figures 14 and 15). These phenomena are verified for both refrigerant flow rates $\dot{m} = 0.003$ kg.s^{-1} and $\dot{m} = 0.004$ kg.s^{-1}.

At given temperatures Tc with 240 K < T_C < 255 K, and for low mass flow rates, $\dot{m} < 0.10$, the trend in the evolution of the Bejan number $Be|_{Ns}$ shows the dominance of heat transfer over viscous friction and $Be|_{Ns} \to 1$. Experimental tests with $\dot{m} = 0.003$ kg.s^{-1} and $\dot{m} = 0.004$ kg.s^{-1} are thus found in this working area of the refrigerating machine.

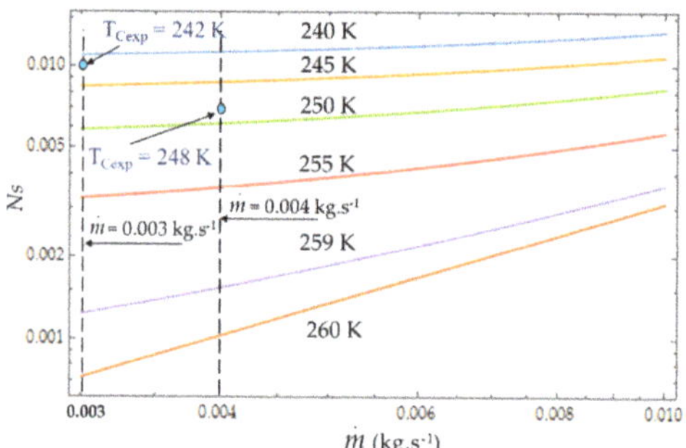

Figure 13. Variations in entropy generation rate N_S as function of mass flow $\dot{m}$ for different temperatures T_C Experimental results for $D_h = 0.015$ m, $L = 0.1$ m, $\dot{m} = 0.003$ kg.s^{-1} and $\dot{m} = 0.004$ kg.s^{-1}.

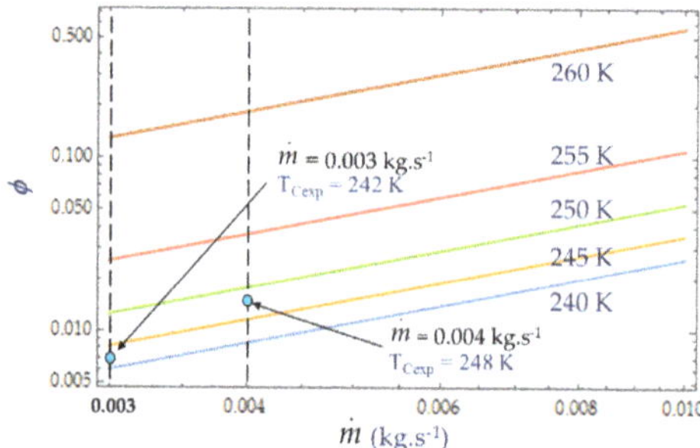

Figure 14. Variation in the irreversibility distribution ratio φ as function of mass flow $\dot{m}$ for different temperatures T_C. Experimental results for $D_h = 0.015$ m, $L = 0.1$ m, $\dot{m} = 0.003$ kg.s^{-1} and $\dot{m} = 0.004$ kg.s^{-1}.

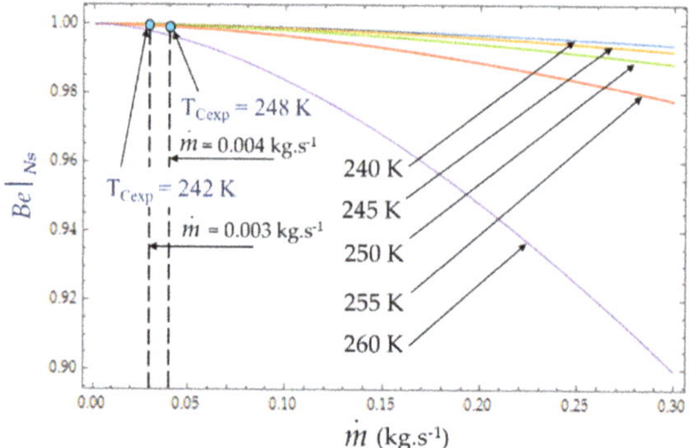

Figure 15. Variation in Bejan number $Be|_{Ns}$ as function of mass flow $\dot{m}$ for different temperatures T_C. Experimental results for $D_h = 0.015$ m, $L = 0.1$ m, $\dot{m} = 0.003$ kg.s^{-1} and $\dot{m} = 0.004$ kg.s^{-1}.

4.3. Effect of Cold Gas Temperature T_C

The temperature T_C of the cold gas inside the expansion volume is measured when the Stirling refrigerator runs at a steady rotational speed ω. For given inlet and outlet temperatures, $T_i = 300$ K and $T_o = 260$ K respectively, the heat transfer exchanged with the collant at mean bulk temperature decreases with the coolant temperature (Figure 16). This figure shows that, at the given cold gas temperature Tc, the entropy generation rate N_S increases with the length L of the heat exchanger due to an increased pressure drop caused by viscous friction along the heat exchanger confirming the dominance of pressure drop [29,42]. As the amplitudes of gas temperature increase, both heat transfer exchanged with the coolant fluid at mean bulk temperature and irreversibility distribution ratio φ increase (Figure 17). At low temperatures (e.g., 240 K), the viscosity and the Prandlt number of the liquid increase sharply, with a factor around 7 between 243 K and 263 K, and the friction losses become high even at low flow rates. For the limit temperature $Tc \to 260$ K, at fixed mass flow $\dot{m}$, the heat transfer would require an infinite quantity or a perfect heat exchanger efficiency and that is why the irreversibility distribution ratio φ rises sharply as the temperature approaches the limit $Tc \to 260$ K and would tend to infinity for $Tc = 260$ K.

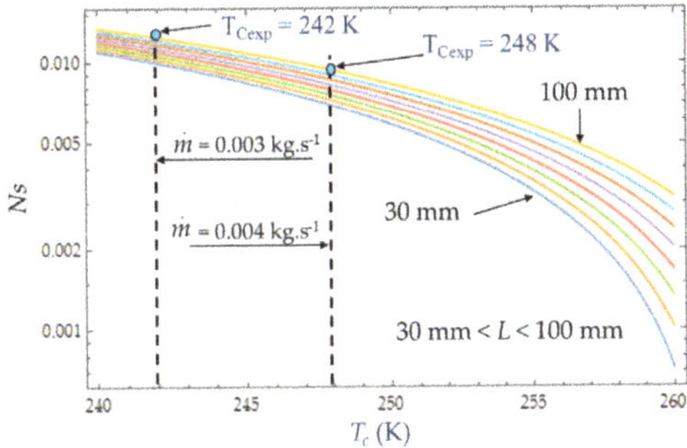

Figure 16. Variations in entropy generation rate N_S as function of temperature T_C for different lengths L. Experimental results for $D_h = 0.015$ m, $L = 0.1$ m, $\dot{m} = 0.003$ kg.s⁻¹ and $\dot{m} = 0.004$ kg.s⁻¹.

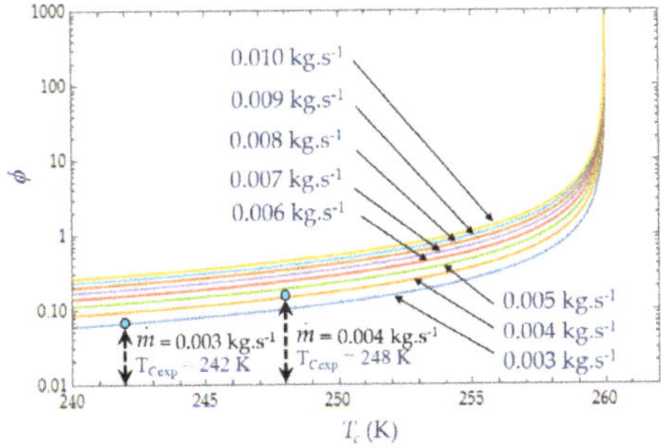

Figure 17. Variation in the irreversibility distribution ratio φ as function of temperature Tc for different mass flow $\dot{m}$. Experimental results for $D_h = 0.015$ m, $L = 0.1$ m, $\dot{m} = 0.003$ kg.s⁻¹ and $\dot{m} = 0.004$ kg.s⁻¹.

5. Conclusions

We have developed an analysis of the entropy generation rate inside a cold heat exchanger of a Stirling refrigerator by examining the behavior of three criteria: the adimensionless entropy ratio N_S, the irreversibility distribution ratio φ and the Bejan number $Be|_{Ns}$ based on the adimensionless entropy ratio $N_{S,\Delta T}$ and $N_{S,\Delta P}$ corresponding to the heat transfer and fluid friction irreversibilities, respectively.

The cold heat exchanger is a double-pipe type heat exchanger made up of two concentric circular tubes. The incompressible coolant liquid flows continuously, with a mass flow rate, through the annular passage, under the temperature and pressure gradients, respectively. The other fluid is the working gas (Nitrogen) of the Stirling refrigerator and it flows alternatively (corresponding to the rotational frequency of the machine) at constant temperature T_C in the inner tube corresponding to the cylinder of the machine.

The cold heat exchanger transfers heat from a gas to a coolant liquid in the temperature range (240–300 K) and the efficiency of this operation is a result of the competition between heat transfer and fluid flow irreversibilities. We have carried out experiments with two mass flow rates $\dot{m} = 0.003$ kg.s⁻¹

and $\dot{m} = 0.004$ kg.s^{-1} and showed the dominance of heat transfer over viscous friction in the cold heat exchanger, which presents a hydraulic diameter $D_h = 0.015$ m and a length $L = 0.10$ m.

The aim of the analysis was to understand the impact of hydraulic diameter D_h, length L, coolant mass flow $\dot{m}$ and cold gas temperature T_C on these three criteria. It could be shown that, to ensure a minimal entropy production ratio N_S, it is necessary to minimize the fluid friction irreversibilities ($N_{S,\Delta P}$). Experimental results have shown the trade-off between fluid flow irreversibilities and heat transfer. We showed that large characteristic length and small hydraulic diameter generate large entropy production especially at low mean temperatures because of the high value of the coolant liquid (mix water/ethylene–glycol) viscosity increasing the fluid friction.

Our analysis needs to be further improved on these different points:

- Develop a second law analysis based on a heat exchanger submitted to periodic flows;
- Investigate other cold exchanger geometries with a larger hydraulic diameter and length and different mass flow rates;
- Perform exergetic and thermoeconomic analysis in order to define new indicators.

Author Contributions: Conceptualization, F.L.; methodology, S.D.-G., F.L. and S.B.; validation, S.D.-G., S.B. and F.L.; formal analysis, F.L.; investigation, S.D.-G.; writing—original draft preparation, F.L.; writing—review and editing, S.D.-G., F.L. and S.B. All authors have read and agreed to the published version of the manuscript.

Funding: This work has been supported by the EIPHI Graduate School (contract "ANR-17-EURE-0002").

Conflicts of Interest: The authors declare no conflict of interest.

References

1. IIF-IIR. *The Role of Refrigeration in the Global Economy—38th Informatory Note on Refrigeration Technologies*; Technical Report; International Institute of Refrigeration: Paris, France, 2019.
2. IEA. *The Future of Cooling, Opportunities for Energy-Efficient Air Conditioning*; Technical Report; IEA: Paris, France, 2018; Available online: https://www.iea.org/futureofcooling/ (accessed on 13 February 2020).
3. Hargreaves, C.M. *The Phillips Stirling Engine*; Elsevier Science Pub. Co., Inc.: New York, NY, USA, 1991.
4. Song, C.F.; Kitamura, Y.; Li, S. Evaluation of Stirling cooler system for cryogenic CO_2 capture. *Appl. Energy* **2012**, *98*, 491–501. [CrossRef]
5. Walker, G. *Cryocoolers—Part 1: Fundamentals*; Springer: Berlin/Heidelberg, Germany, 2014.
6. Cheng, C.H.; Huang, C.Y.; Yang, H.S. Development of a 90-k beta type Stirling cooler with rhombic drive mechanism. *Int. J. Refrig.* **2019**, *98*, 388–398. [CrossRef]
7. Hewitt, G.F. *Heat Exchanger Design Handbook*; Begell House: New York, NY, USA, 1992.
8. Hesselgreaves, J.E. *Compact_Heat_Exchangers_Selection Design and Operation*; Butterworth-Heinemann: Pergamon, Turkey, 2001.
9. Kuppan, T. *Heat Exchanger Design Handbook*; Marcel Dekker: New York, NY, USA, 2000.
10. Shah, R.K.; Sekulic, D.P. *Fundamental of Heat Exchanger Design*; Wiley: New York, NY, USA, 2003.
11. Walker, G. *Stirling Engines*; Clarendon Press; Oxford University Press: Oxford, UK, 1980.
12. Walker, G.; Senft, J.R. *Free Piston Stirling Engine*; Springer: Berlin, Heidelberg, 1985.
13. Sciubba, E. On the second-law inconsistency of emergy analysis. *Energy* **2010**, *35*, 3696–3706. [CrossRef]
14. Lucia, U. Entropy and exergy in irreversible renewable energy systems. *Renew. Sustain. Energy Rev.* **2013**, *20*, 559–564. [CrossRef]
15. Lucia, U.; Grisolia, G. Exergy inefficiency: An indicator for sustainable development analysis. *Energy Rep.* **2019**, *5*, 62–69. [CrossRef]
16. Bejan, A. General criterion for rating heat-exchanger performance. *Int. J. Heat Mass Transf.* **1978**, *21*, 655–658. [CrossRef]
17. Poulikakos, D.; Bejan, A. Fin geometry for minimum entropy generation in forced convection. *J. Heat Transf.* **1982**, *104*, 616–623. [CrossRef]
18. Bejan, A. The thermodynamic design of heat and mass transfer processes and devices. *Int. J. Heat Fluid Flow* **1987**, *8*, 258–276. [CrossRef]

19. Sciacovelli, A.; Verda, V.; Sciubba, E. Entropy generation analysis as a design tool—A review. *Renew. Sustain. Energy Rev.* **2015**, *43*, 1167–1181. [CrossRef]

20. Nag, P.K.; Kumar, N. Second law optimization of convective heat transfer through a duct with constant heat flux. *Int. J. Energy Res.* **1989**, *13*, 537–543. [CrossRef]

21. Sahin, A.Z. Irreversibilities in various duct geometries with constant wall heat flux and laminar flow. *Energy* **1998**, *23*, 465–473. [CrossRef]

22. Zimparov, V. Extended performance evaluation criteria for enhanced heat transfer surfaces: Heat transfer through ducts with constant wall temperature. *Int. J. Heat Mass Transf.* **2000**, *43*, 3137–3155. [CrossRef]

23. Zimparov, V. Extended performance evaluation criteria for enhanced heat transfer surfaces: Heat transfer through ducts with constant heat flux. *Int. J. Heat Mass Transf.* **2001**, *44*, 169–180. [CrossRef]

24. Khan, W.A.; Culham, R.J.; Yovanovich, M.M. Fluid flow around and heat transfer from elliptical cylinders: Analytical approach. *J. Thermophys. Heat Transf.* **2005**, *19*, 178–185. [CrossRef]

25. Begot, S.; Dejtel-Gothe, S.; Khirzada, H.; Lanzetta, F.; Layes, G.; Nika, P. Machine Stirling de Type Beta. Patent FR 1873540/WOBR18SRERPV/FH, 19 March 2018.

26. Yilmaz, M.; Sara, O.N.; Karsli, S. Performance evaluation criteria for heat exchangers based on second law analysis. *Exergy Int. J.* **2001**, *1*, 278–294. [CrossRef]

27. Bejan, A. Second law analysis in heat transfer. *Energy* **1980**, *5*, 720–732. [CrossRef]

28. Rosen, M.A. Second-law analysis: Approaches and implications. *Int. J. Energy Res.* **1999**, *23*, 415–429. [CrossRef]

29. Bejan, A. *Entropy Generation Minimization: The Method of Thermodynamic Optimization of Finite-Size Systems and Finite-Time Processes*; CRC Press: Boca Raton, FL, USA, 2013.

30. Natalini, G.; Sciubba, E. Minimization of the local rates of entropy production in the design of air-cooled gas turbine blades. *J. Eng. Gas Turbines Power* **1999**, *121*, 466–475. [CrossRef]

31. Sciubba, E. Optimisation of turbomachinery components by constrained minimisation of the local entropy production rate. In *Thermodynamic Optimization of Complex Energy Systems*; Springer: Berlin/Heidelberg, Germany, 1999; pp. 163–186.

32. Kagawa, N. *Regenerative Thermal Machines (Stirling and Vuilleumier Cycle Machines) for Heating and Cooling*; International Institut of Refrigeration: Paris, France, 2000.

33. Zimparov, V.D.; Vulchanov, N.L. Performance evaluation criteria for enhanced heat transfer surfaces. *Int. J. Heat Mass Transf.* **1994**, *37*, 1807–1816. [CrossRef]

34. Lotfi, B.; Sundén, B. Development of new finned tube heat exchanger: Innovative tube-bank design and thermohydraulic performance. *Heat Transf. Eng.* **2019**, 1–23. [CrossRef]

35. Li, M.; Lai, A.C.K. Thermodynamic optimization of ground heat exchangers with single u-tube by entropy generation minimization method. *Energy Convers. Manag.* **2013**, *65*, 133–139. [CrossRef]

36. Bejan, A. *Convection Heat Transfer*; John Wiley & Sons: Hoboken, NJ, USA, 2013.

37. Kandlikar, S.; Garimella, S.; Li, D.; Colin, S.; King, M.R. *Heat Transfer and Fluid Flow in Minichannels and Microchannels*; Elsevier: Amsterdam, The Netherlands, 2005.

38. Djetel-Gothe, S. Modélisation et Réalisation d'une Machine Réceptrice de Stirling Pour la Production de Froid. Ph.D. Thesis, University Bourgogne Franche-Comté, Belfort, France, 2020.

39. L&B. Available online: http://www.protekt.ch/ (accessed on 13 February 2020).

40. Sahin, A.Z. Entropy generation in turbulent liquid flow through a smooth duct subjected to constant wall temperature. *Int. J. Heat Mass Transf.* **2000**, *43*, 1469–1478. [CrossRef]

41. Khan, W. Modeling of Fluid Flow and Heat Transfer for Optimization of Pin-Fin Heat Sinks. Ph.D. Thesis, University of Waterloo, Waterloo, ON, Canada, 2004.

42. Sahiti, N.; Krasniqi, F.; Fejzullahu, X.H.; Bunjaku, J.; Muriqi, A. Entropy generation minimization of a double-pipe pin fin heat exchanger. *Appl. Therm. Eng.* **2008**, *28*, 2337–2344. [CrossRef]

MDPI

St. Alban-Anlage 66

4052 Basel

Switzerland

Tel. +41 61 683 77 34

Fax +41 61 302 89 18

www.mdpi.com

Entropy Editorial Office

E-mail: entropy@mdpi.com

www.mdpi.com/journal/entropy